Policy Making in an Era of Global Environmental Change

Environment & Policy
Volume 6

# Policy Making in an Era of Global Environmental Change

Edited by

R. E. MUNN,
J. W. M. la RIVIÈRE
and
N. VAN LOOKEREN CAMPAGNE

**KLUWER ACADEMIC PUBLISHERS**
DORDRECHT / BOSTON / LONDON

*Disclaimer*: The views of individual contributors (chapter authors and interviewees) as expressed in their contributions are not necessarily shared by the editors and other contributors.

**Library of Congress Cataloging-in-Publication Data**
Policy making in an era of global environmental change / edited by
R. E. Munn and J. W. M. la Rivière and N. van Lookeren Campagne.
p. cm. — (Environment & policy; v. 6)
Includes index.
ISBN 0–7923–4072–8 (acid free paper)
1. Global environmental change — Government policy. I. Munn, R. E.
II. La Rivière, J. W. M. III. Lookeren Campagne, N. van. IV. Series.
GE149.P65 1996
363.7'0056—dc20 96–17818

---

Published by Kluwer Academic Publishers,
P. O. Box 17, 3300 AA Dordrecht, The Netherlands.

Kluwer Academic Publishers incorporates
the publishing programmes of
D. Reidel, Martinus Nijhoff, Dr W. Junk and MTP Press.

Sold and distributed in the U.S.A. and Canada
by Kluwer Academic Publishers,
101 Philip Drive, Norwell, MA 02061, U.S.A.

In all other countries, sold and distributed
by Kluwer Academic Publishers Group,
P. O. Box 322, 3300 AH Dordrecht, The Netherlands.

*Printed on acid-free paper*



**Printed and bound in Great Britain by Hartnolls Limited, Bodmin, Cornwall**

# Contents

# Preface by the Sponsors

The preparation of this book was initiated by a Steering Committee of Chief Executive Officers of major Dutch industries and some scientists. It was set up in 1991 and was joined by six Dutch institutions interested in the science/society interface. Sixteen industries provided funding and many individuals assisted the venture by support in kind.

## Purpose, contents and target group

Major international, interdisciplinary research programmes are underway to increase our understanding of how the Earth System operates and how it is changing. This research involves the complex physical, chemical and biological interactions between atmosphere, oceans and continents which regulate the unique environment that the Earth System provides for life. It involves the changes – natural and man-made – in the system and the manner in which these will affect society. The international research effort is unprecedented in scale and in the cooperation between scientists and (inter)governmental bodies.

Although the understanding of the Earth System, and the prediction of its carrying capacity, are still limited, scientists already agree that significant changes in global and regional environments can be expected in the next 50 years and that these will affect the possibilities of the Earth to sustain life. Insight into the relations between the pressures on the Earth System and its carrying capacity are a prerequisite for planning a more sustainable future.

Stakeholders in the global change issue and the ongoing associated research are: (1) governance at national, regional and international levels; (2) business/industry and the consumers of their products, and (3) other non-governmental interest groups. In the development of societal responses to environmental changes, it is vital that policy makers and scientists understand one another so as to make effective cooperation possible. This book aims to provide a contribution to it.

The first part of the book gives a description of what is understood by global change and presents an overview of the ongoing relevant research, focusing on two major research programmes described by its scientific leaders in an easily understandable and entertaining way. In the second part an analytical overview is given of the response process exemplified by responses at the intergovernmental and governmental level, and by business and public interest groups.

Thus for the first time one book describes both ongoing research work in global change and the response processes that the research results evoke. It is of interest to the decision makers of these stakeholders. They are the target group for which the book has been written.

It is hoped that the book will assist in the process of increasing cooperation between policy makers and the scientific community.

Some of the elements in this interaction brought out in this book are:

- The international research framework described in the first part of this book helps national research programmes to find their place, so as to avoid duplication, to save funds and manpower and to promote synergy. In this planning and coordination and also in the implementation of the programmes and the synthesizing of results, the independent, non-governmental International Council of Scientific Unions (ICSU) plays an important role. It also helps to ensure scientific quality of the research and especially scientific objectivity. It painstakingly adheres to non-biased fact-finding and abstains from speculative interpretation of results. It also helps to safeguard the programmes from outside pressures, while maintaining an open ear to priority needs voiced by policy makers. The Steering Committee for this monograph has been fortunate to receive the full cooperation of ICSU and of the leaders of the programmes it is involved in. The overviews of ongoing research provided in the book should facilitate access by the reader to the primary sources of Earth System research and thus help in separating facts from fiction and placing these facts in a wider context.
- In planning the construction of major installations with a lifetime of some 40 years or longer (e.g. refineries, electricity supply, mines, dams) relevant global change issues should be taken into account. Keeping track of the research issues as they are produced can help priority setting in planning and running the installations and can provide an early warning of enhanced risks, locally and regionally. The book describes the emerging methodologies to do this.
- Global change issues will affect the market place in many ways, involving e.g. agriculture, forestry, insurance and the development of energy sources. Besides new risks, new business opportunities also arise in sectors as diverse as water resource technology, air conditioning, biotechnology and dredging harbours or building coastal works. It is therefore useful to follow research results closely as well as the responses that are taking shape in international conventions and assessment structures, to which the book provides an introduction.
- The new and complex aggregate of response mechanisms to expected global changes is still developing and it is for obvious reasons desirable that major stakeholders participate in this process. This book attempts to assist especially the decision makers in business and industry to join on a pro-active basis those in (inter)governmental and science organizations.

It is realized that this book appears early in the day, as we stand at the beginning of a development in which insight into global change processes is just beginning to be incorporated into the long-term thinking of many decision makers in society.

Already in 1992 the Steering Committee had organized a symposium by scientists for decision makers in the Netherlands as a first step in promoting mutual understanding with respect to global change issues and research. Secretarial work was conducted by the staff of the Royal Institution of Engineers in the Netherlands (KIvI). This book is a second step in the same direction in which advantage has been taken of the experiences gained in the symposium and the progress in the intervening years in integrating and developing global change research.

## Acknowledgments

The Steering Committee and the six participating Institutions are most grateful to the 16 internationally operating Dutch industries* that financially assisted the 1992 Symposium and the preparation of this book:

- The oil, gas and chemical company, SHELL NEDERLAND BV
- The royal dutch steelmills, KONINKLIJKE HOOGOVENS NV
- The Netherlands Organization for Applied Scientific Research, TNO
- The Netherlands Energy Research Foundation, ECN
- The chemicals & materials company, DSM NV
- The international company, AKZO NOBEL NV
- The agricultural and general AAA-bank, RABO BANK NED. BV
- The starch and derivation producers, AVEBE BA.
- The Information Technology Services Group, BSO/ORIGIN BV
- The rubber and tyre company, VREDESTEIN N. V.
- The distribution, packaging and paper group, KONINKLIJKE KNP BT.
- The royal transportation group, KONINKLIJKE NEDDLOYD N. V.
- The dutch chemical company, HOECHST HOLLAND BV
- The chemical company, GE PLASTICS EUROPE
- The Foundation 'Central Institute for Industry', CIVI
- NV SEP/Dutch Electricity Generating Board.

The six supporting Dutch Institutions are:

- The Society and Enterprise Foundation (1968) (Stichting Maatschappij en Onderneming; SMO)
- The Dutch Society of Sciences (1752) (Hollandsche Maatschappij der Wetenschappen; HMW)

*More information about these industries is given in appendix A.1.

- The Royal Society for Agricultural Sciences (1886) (Koninklijk Genootschap voor Landbouwwetenschap; KGvL)
- The Royal Institution of Engineers in the Netherlands (1847) (Koninklijk Instituut van Ingenieurs; KIvI)
- The Royal Netherlands Academy of Arts and Sciences (1808) (Koninklijke Nederlandse Academie van Wetenschappen; KNAW)
- The Netherlands Society for Industry and Trade (1777) (Nederlandsche Maatschappij voor Nijverheid en Handel; NMNH)

The members of the Steering Committee of this endeavour are:

Prof. C. J. F. Böttcher
P. van Duursen
N. G. Ketting
Prof. H. H. van de Kroonenberg
N. van Lookeren Campagne
F. C. Rauwenhoff
Prof. J. W. M. la Rivière
O. H. A. van Royen
C. W. van der Wal

The Chairman of the Steering Committee

(N. van Lookeren Campagne)

# Executive Summary

Will major global environmental change take place in the 21st century? How much is known about the change? Can we find out more? Must we take it seriously? Should we do something now, or can we wait a decade or so? This book seeks to answer these questions.

Scientists are reaching consensus that rates of change in many of the environmental indicators are already increasing in some cases, and are likely to increase even more rapidly in the 21st century. Although there are still uncertainties, the case is made that the consequences of not taking action are too great to take a chance that the scientists' expectations are wrong. The book leads to the admonition – Think long term, act now!

## Signs of Change (Chapter 1)

Chapter 1 provides a definition of global change and some examples of changes already taking place in the driving forces (demographics, energy consumption, deforestation, 'mining of groundwater', etc.) and in some of the environmental variables (increasing atmospheric carbon dioxide concentrations, depletion of stratospheric ozone, desertification, decreasing biodiversity, etc.) The chapter raises some important questions for science and society, and concludes that global environmental change is both a scientific and a political issue.

## Research in Progress (Chapters 2–4)

Chapter 2 gives an overview of the main international research programmes aimed at increasing our understanding of how the Earth System operates and of how it is likely to change in future decades. The point is emphasized that in order to improve our predictive capacity, improved understanding of the Earth System is a prerequisite.

Only in recent years has a comprehensive study of the Earth System been possible, with the development of complex satellite instrumentation for observing the Earth, advanced global telecommunication and computer systems, and greatly expanded numerical models. The sum total of ongoing research is unprecedented in size and complexity, calling for new levels of international cooperation among scientists and between scientists and governmental and intergovernmental bodies, as well as business and industry.

Chapter 3 describes the International Geosphere Biosphere Programme (IGBP) launched by the International Council of Scientific Unions (ICSU). IGBP ambitiously addresses the interactive physical, chemical and biological systems that regulate the total Earth System, the unique environment that this system provides for life, the changes occurring in the system and the manner in which these changes are influenced by human activities. In order to be manageable, IGBP focuses its work on a limited number of high priority 'core' issues. The rationale for assigning these priorities is explained. One of the projects, for example, examines global changes of the distant past – as documented by fossils, sediments, pollen deposits, ice cores and tree rings – in order to glean some clues about the future. Other projects focus on atmospheric chemistry, land–ocean interactions in coastal zones, biospheric aspects of the water cycle, changes in land cover and land use, terrestrial ecosystems and ocean/atmosphere flows of trace gases.

Chapter 4 outlines the climate system – the atmosphere, oceans, ice surfaces and land surfaces. The natural 'greenhouse' effect is discussed, as well as its enhancement by gases such as carbon dioxide generated by human activities. Most of the ongoing global research is carried out under the auspices of the World Climate Research Programme (WCRP), sponsored by the World Meteorological Organization (WMO), the International Council of Scientific Unions (ICSU) and the Intergovernmental Oceanographic Commission (IOC). The objectives are to determine to what extent climate (in contrast to weather) can be predicted and how human activities affect it, which requires differentiation of natural variability from human-induced change. WCRP's work is organized into six projects, one of which – TOGA (the Tropical Ocean and Global Atmosphere study – has already had remarkable results in successfully predicting interannual climate variations connected with the so-called El-Niño phenomenon, helping farmers in some regions to decide what crops to plant. Other projects deal with topics like climate variability and predictability, the global energy and water cycles, and the world ocean circulation.

WCRP and IGBP maintain close liaison in order to avoid duplication and to achieve synergies, WCRP focusing on the physical aspects of climate change, and IGBP taking care of the chemical and biological aspects.

## Policy Responses to Global Environmental Issues (Chapters 5–9)

Chapter 5 provides an introductory overview of the policy responses that have been taken by intergovernmental and governmental bodies and by business and environmental non-governmental organizations over the last 15 years. This historical review shows that formulation and execution of policy responses is a gradual process, still evolving but already marked by a number of important intergovernmental events – stretching from Stockholm to Rio.

1972: The UN Stockholm Conference on the Human Environment
1983: Establishment of the UN 'Brundtland' World Commission on Environment and Development
1987: Establishment of the Intergovernmental Panel on Climate Change (IPCC)
1990: Second World Climate Conference and Ministerial Sessions
1992: The U N Rio Conference on Environment and Development (UNCED).

Local pollution issues evoke immediate policy responses and thus lead easily to acceptable regulatory legislation. In contrast, it is difficult to persuade citizens, in all parts of the world, that action must be taken now to resolve some of the global environmental issues of the 21st century, which are:

(1) global, requiring world-wide concerted actions (national efforts on energy conservation, for example, are irrelevant if not adopted in many countries.)
(2) long-term (The Brundtland idea of the need for sustainable futures is slowly being adopted.)
(3) shrouded in uncertainty (The uncertainty about size and timing of future global changes raises the question of whether action should be postponed until more confidence can be given to the predictions.)

The main *policy* responses so far have been at the intergovernmental level, with independent science bodies such as ICSU providing information and advice. Because of the uncertainties involved, the policies adopted are predicated on the principles of: (1) keeping options open as long as possible; (2) taking precautionary measures if the long-term consequences of not taking action are unacceptable; and (3) adopting a no-regrets approach, i.e. taking actions which can be justified whether the unacceptable 'future' occurs or not. The no-regrets approach is a first defence against unacceptable change.

Chapters 6 and 7 elaborate some of these ideas at the inter-governmental and governmental levels, respectively. Three particular aspects are stressed:

(1) The need for increased understanding of the Earth System, through continuing support for IGBP and related programmes. Some brief descriptions of the national global change research programmes of the Netherlands, Germany, USA, Canada, Japan and Thailand are provided as examples.
(2) The need for improved early warning global monitoring systems, permitting policy revisions as new information is received.
(3) The need for national, regional and global environmental assessments, written specifically for policy makers; the case of climate change is used as an example. The IPCC has performed an important role, producing assessments of current knowledge in a form that is useful for policy makers.

Chapter 8 deals with the policy responses to global environment change that are being taken or being considered by business and industry. No global change action plan can hope to succeed without the active cooperation of business and industry, which represent, together with consumer demand, the driving engine of economic growth. Therefore interaction between global change research and business/industry is essential. Although the greening of industry is making progress with respect to local and regional issues, new initiatives are required for resolving long-term global issues. This chapter concludes with the record of interviews with six senior members of the business and government communities in the Netherlands regarding their views on global environmental change. Although the questions posed to each person were more or less the same, the answers are quite different. Reading the interviews is a refreshing experience.

Finally Chapter 9 deals briefly with environmental non-governmental organizations (NGOs) in the context of global environmental change. Public understanding of the many issues surrounding 'global change' is indispensible, and environmental NGOs make an important contribution to this task. In order to fulfil this role, NGOs require good documentation, which means that they are important 'clients' of the results coming from IGBP, WCRP and other international environmental research programmes.

## Appendices

The monograph concludes with five Appendices, providing information on contact addresses on authors, and supporting and sponsoring organizations; a list of acronyms; and three brief essays describing (a): methods for assessing effects of global change on the biosphere and society; (b) policy instruments for consensus-building and long-term priority setting with respect to global change issues; and (c) adaptation strategies relating to climate change, as developed in the Canadian context; and finally an Index.

## Conclusion

The main thesis of this book is that the possibility of substantial global change in the 21st century is sufficiently high that actions should be taken now:

(1) to improve understanding of the Earth System;
(2) to expand early warning monitoring systems; and
(3) to develop robust 'adaptive' environmental policies.

In order to do this, greater collaboration should be fostered amongst scientists, governmental and intergovernmental bodies, business and industry, and environmental non-governmental bodies.

# Foreword

The future becomes more and more uncertain as time horizons lengthen. There is no doubt, however, that in the year 2050, say, the world will be quite different – socio-economically, politically and environmentally – than it is today. There are two reasons for making this assertion. In the first place, the evidence is everywhere that the world has changed greatly in the last 50 years; some examples in Chapter 1 will illustrate this statement. Secondly, rates of change of a number of environmental indicators have become large, and many scientists predict that they will be unprecedented by the middle of the 21st century. It is prudent to take a risk-assessment approach to managing our responses to these long-term changes. In this connection, it should be mentioned that although the year 2050 seems a long way into the future:

(1) Mega-development projects under construction today are designed to operate for at least 50 years, and in some cases are permanent commitments of land (e.g., urban expansion, large dams, super-highways);
(2) Children born in the 1990s will only be in their fifties in the year 2050. In this perspective, the future is fast approaching.

Fortunately, many scientists and policy people are trying to think 'long-term', by improving understanding of the socioeconomic and environmental forces that shape the future, and by developing policies that minimize risks and exploit opportunities that may occur in a future that will include great change and many surprises. These endeavours are generally lumped together in a field called Global Change, which has attracted the interest of United Nations and major non-governmental bodies, and has led to the development of a number of international research programmes. Two of the most important of these are IGBP and the WCRP (see Chapters 2, 3 and 4).

Historians who reflect back on the Austro-Hungarian Empire find the first signs of its collapse as early as 1850. The factors that caused the subsequent decline of this magnificent kingdom were economic, cultural and political. By 1900, the stresses had become enormous, and Franz-Joseph's Empire was 'waiting for an accident to happen'. That accident occurred in 1914, the assassination of Franz Ferdinand, triggering the start of World War I.

Suppose that Franz Joseph had recruited a group of policy analysts in the 1850s. Could they have provided policy advice over the next 50 years that would have permitted the Empire to remain secure? Recognizing that the future is very uncertain, particularly insofar as occurrences of triggering shocks are concerned, what would this advice have been? That retrospective look at history leads to the question of how today's business, industry and governments can best prepare for a changing future.

The purpose of this book is: *firstly*, to draw attention to the likelihood of unprecedented global environmental change in the 21st century; *secondly*, to discuss the efforts of the international scientific community to improve the projections (at the same time emphasizing that global change is a risk management issue, and that the 'precautionary principle' must prevail); and *lastly*, to discuss the effects of global change on long-term policy formulation in business, industry and government. A positive view is taken, emphasizing the opportunities that exist for the development of fruitful partnerships amongst industry, business, governments and scientists to secure a sustainable future for the globe. At the same time, the book recognizes that serious regional problems may occur, due for example to 'environmental refugees', severe water shortages and disruptions of traditional trade patterns. As a matter of fact, environment is already a brake on development in some parts of the world. The global changes possible in the 21st century could exacerbate such problem areas, as well as widen their extent.

The emphasis in the book will be on the urgency of including global environmental change in economic policy formulation: 'Think long-term – act now!'

This book is written for senior officials (particularly policy analysts) in business, industry and government (intergovernmental, national, provincial and city levels). Every effort is made to avoid technical terms and jargon, although some acronyms will be used. The book has been written insofar as possible to appeal to non-specialists interested in the future of this Planet, who nevertheless do not wish to read yet some more Doomsday predictions.

Finally it should be mentioned that the monograph will not discuss North–South issues, the fundamental driver for the Brundtland Report and the Rio UNCED Conference. That would broaden the subject matter too greatly, although a global perspective of the biogeophysical world must surely be taken.

**Acronyms of Special Importance in this Monograph**

ICSU: International Council of Scientific Unions
IGBP: International Geosphere–Biosphere Programme
IHDP International Human Dimensions of Global Environmental Change Programme
IPCC: Intergovernmental Panel on Climate Change
UNEP: United Nations Environment Programme
WCRP: World Climate Research Programme
WMO: World Meteorological Organization

Note: A more complete list of acronyms is given in Appendix 2.

# 1. Global Change: Both a Scientific and a Political Issue

R. E. MUNN
*Institute for Environmental Studies, University of Toronto*

## Introduction: What is Global Change?

'Global change' is a phrase introduced by natural scientists in the early 1980s, and later used widely by social scientists, policy analysts, engineers and environmental action groups. The phrase has quite a specific connotation, expressing a concern that the world is changing faster than ever before. 'Global change' incorporates an anxiety that by the late 21st century, large-scale demographic, socioeconomic, technological, cultural and environmental conditions will be greatly different from those prevailing today – and probably not at all to our liking. The fact that many of the changes will be global in scale makes the situation even more worrisome. Nation states will not be able to protect themselves individually from harmful effects.

The world community of scientists has responded to these concerns in various ways. Most importantly, ICSU (the International Council of Scientific Unions) has launched the international program IGBP (International Geosphere–Biosphere Programme) involving scientists from different disciplines and countries. The principal objective of IGBP is to improve our understanding of the land–ocean–atmosphere system – an urgent goal if society is to prosper in the 21st century.

**Global change** is a term intended to encompass the full range of global issues and interactions concerning natural and human-induced changes in the Earth's environment. The United States Global Change Research Act of 1990 defines **global change** as 'changes in the global environment (including alterations in climate, land productivity, oceans or other water resources, atmospheric chemistry and ecological systems) that may alter the capacity of the Earth to sustain life.'

From *Our Changing Planet, the FY95 US Global Change Research Program*, p. 2 (a report of the United States National Science and Technology Council, Washington, 1994)

The natural environment is continually changing, but mostly on time scales of millennia. The additional driving forces of major concern today are human population pres-

*R. E. Munn, J. W. M. la Rivière and N. van Lookeren Campagne (eds), Policy Making in an Era of Global Environmental Change, 1–15.* 

sures, overconsumption of natural resources, and overproduction of wastes. In order to understand these processes, and to design appropriate response strategies, contributions from social as well as natural sciences are required.

In the following sections of this introductory overview, examples will be given of recent observed 'global' trends, including projections over the next two decades where current trends are practically unstoppable over the mid-term, and of some possible future trends based on simulation model calculations, in both socioeconomic and environmental indicators. This will lead into a discussion of the consequent critical issues for science and society.

## Examples of Some Recent 'Global' Trends

In the last century, mankind's influence on the biosphere was local and immediate. In recent years, however, the impacts of society on the environment have become increasingly large-scale, long-term and in some cases almost irreversible. Examples of these new kinds of issues are acidic deposition, the Antarctic stratospheric ozone hole, deforestation, desertification and losses of genetic resources. As this trend accelerates in the 21st century, the need to understand the Earth's biogeophysical systems, including the driving forces, will become more and more urgent.

The sequence of the examples given below does not indicate priorities.

### *Rising Numbers of People and Automobiles*

Figure 1 compares on a relative scale the world's populations of people and of passenger cars from 1950 to the present. Both show dramatic increases but the number of motor

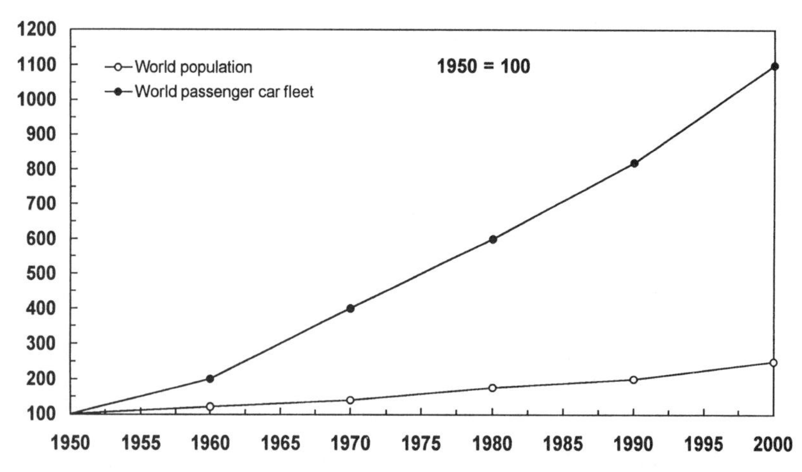


Fig. 1. Increases in numbers of people and passenger cars in the world between 1950 and 2000 (Walsh, 1990; UNEP, 1992). Relative scales are used: 1950 = 100)

vehicles is going up much faster than of people. Taking a global perspective, the present number of automobiles and trucks (540 million) is expected to exceed 1 billion early in the 21st century (Mackenzie, 1991). The number of automobiles in the UK is expected to increase from 21 million in 1988 to 36 million by the year 2030, with a planned expansion of the British road system to match the demand. Whether this trend can be arrested (through higher taxes on gasoline and increased investment in public transport) remains to be seen. Certainly the number of cars per family is increasing in Britain, and the car occupancy rate is decreasing.

Much attention has been given to demographic trends in developing countries. However, the fact that rapid urban population growth is not only a Third World problem is illustrated in Figure 2, which shows the growth of the Greater Vancouver Regional District (a coastal area in western Canada) from 1921 (291 000) to 1991 (1.7 million), with a projected value of 3 million being reached sometime between 2015 and 2031 (Baxter and Laglagaron, 1993). Although the number of automobiles is increasing faster than the number of people, technology and emission controls have managed to limit total automobile emissions in this urban area. Early in the 21st century, however, emissions will begin to increase again.

Vancouver's growth in population is an example of a world-wide phenomenon. In 1970, 50% of the world's population lived in coastal areas; today the figure is 60% (IUCN/UNEP/WWF, 1991). Thus there is an accelerating loss of prime agricultural land and wetlands, and there are shortages of fresh water, particularly in the Third World. Some cities there are sufficiently stressed that 'they are waiting for an accident to happen'. They have lost the ability to withstand severe shocks such as famines and plagues.

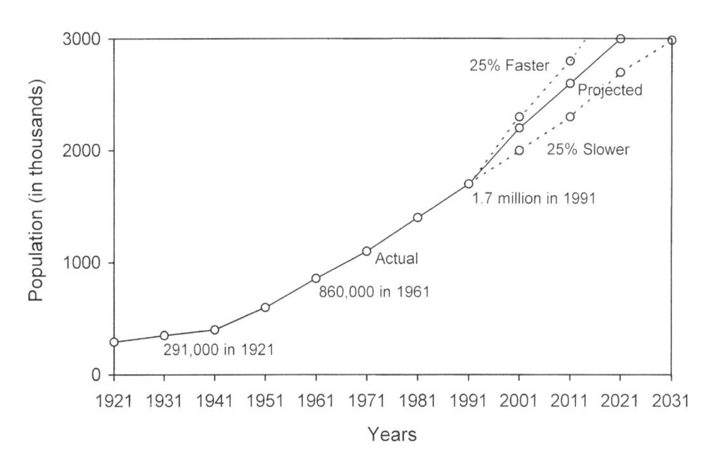


Fig. 2. Population growth in the Metropolitan Vancouver Region, Canada, including extrapolation to year 2031 (Baxter and Langladaron, 1993). This example shows that urban growth is not a problem only for Third World cities.

*The Recent Decline in Lead (Pb) Levels: a success story*

It is appropriate to include a success story, which shows that potentially harmful environmental trends can sometimes be reversed, given sufficient physical understanding of the processes involved, and a determination by society to take the necessary action. This story concerns lead (Pb), which comes mainly from automobiles and foundries, particularly car-battery manufacturing plants. During the last two decades, there has been a steep decline in lead emissions in many industrialized countries due to improved technology and to air pollution control efforts. This has resulted in a steep decline in lead concentrations in cities, and in lead levels in human blood. See for example, the 1990 report by Hillborn and Still on downward trends in blood levels of Pb in Canadian children in urban areas.

Evidence that these trends are also global comes from the analysis of Greenland ice cores, which provide data on annual deposition of lead in snow. Lead emissions are carried by the winds from industrialized areas in the Northern Hemisphere, and as shown in Figure 3, annual Pb deposits in Greenland ice cores have been declining in the 1970s and 1980s following the sharp rise in the 1940s and 1950s. As a matter of fact, lead concentrations have recently been detected in Greenland ice cores that were four times above background levels at the time of the Greek and Roman civilizations (500 BC to 300 AD) (Hong *et al.*, 1994). In those days, considerable quantities of lead and silver were mined and smelted, mostly in the Mediterranean area. Greenland ice cores have also revealed elevated copper deposition during Roman and medieval times (Hong *et al.*, 1996).

Other success stories include: the improvement in water quality in the Rhine and Thames rivers; the recovery of an acidified lake near Sudbury, Canada, once a smelter had closed down (Gunn and Keller, 1990); the reduction in PCB concentrations in the

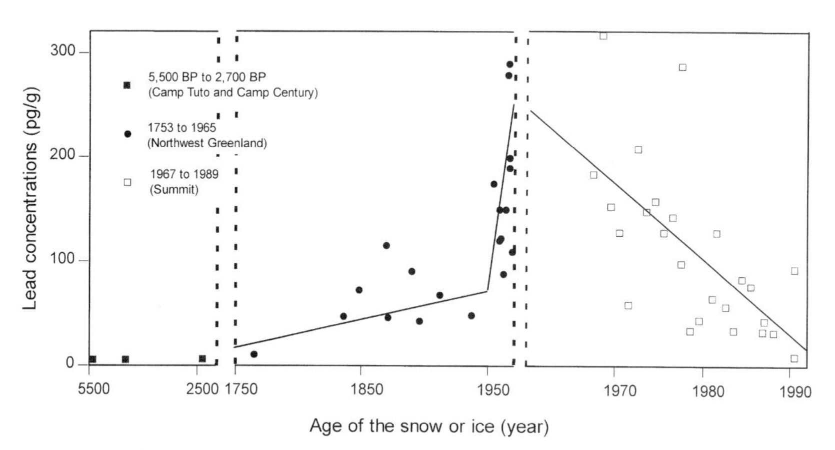


Fig. 3. Changes in lead (Pb) concentrations in Greenland ice and snow cores from 5500 BP (before the present) to the present. The decline in the last 20 years is mainly due to the decreasing use of lead additives in gasoline in Europe and North America (Boutron *et al.*, 1991)

eggs of guillemots feeding on herring in the Baltic; and the virtual elimination of eutrophication in some freshwater lakes following tight regulation of phosphate releases.

These examples illustrate the fact that society can sometimes reverse undesirable environmental changes. Although some parts of the system may seem out of control, the situation is not yet hopeless.

### *Dwindling Fresh Water Supplies*

Figure 4 shows trends in annual *per-capita* availability of fresh water in developing countries. In these continents and in some industrialized countries, fresh water is being 'mined' due to increased *per-capita* water consumption and increased population. In some towns in Northern Nigeria, for example, 15% of family income is spent on purchases of drinking water. See also Postel *et al.* (1996).

River diversions, particularly for irrigation in semi-arid regions, have caused international tension for centuries. Examples are to be found in North Africa and in the Tazhakstan region of Asia, which is almost totally dependent on run-off from the Himalayas. If these regions were to become warmer and drier, some human settlements would not be able to survive, no matter how water efficient the inhabitants became. Fresh-water scarcity in the Middle East is a well-known complicating factor in the political situation there.

Attention has recently been drawn to a related issue – engineering interventions on river systems (construction of large dams, flooding of reservoirs, etc.) (Dynesius and Nilsson, 1994). The number of large dams has increased more than sevenfold since 1950, and the land inundated by reservoirs is comparable with the area of France. Dynesius and Nilsson estimate that 77% of the total water discharge from the 139 major rivers in the northern third of the world is strongly regulated. The consequences for terrestrial and marine ecosystems may be enormous.

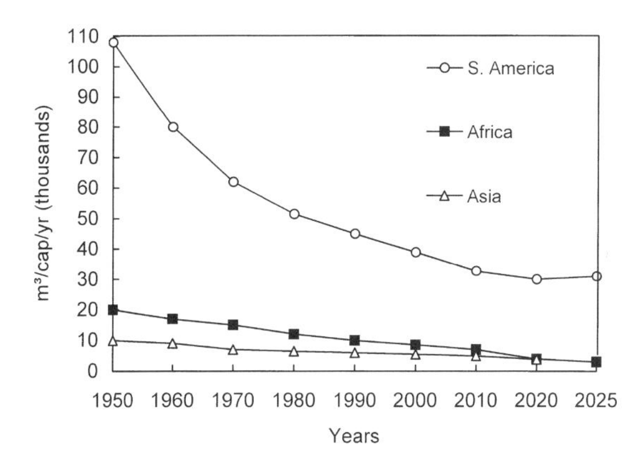


Fig. 4. Decreasing annual *per-capita* availability of fresh water in developing countries (Ayibotele and Falkenmark, 1992)

*The Stratospheric Ozone Hole*

A decrease in stratospheric ozone was first reported by Farman *et al.* (1985) at the Halley Bay Station in Antarctica beginning in the mid-seventies (see Figure 5). Subsequent satellite observations showed an ozone 'hole' over Antarctica, with a decrease of about 30–40% in a decade during the Austral spring (October). For a few years ozone depletion was believed to be purely an Antarctic phenomenon but there is now strong evidence of ozone depletion in the middle and high latitudes of the Northern Hemisphere (Bojkov, 1992).

> Stratospheric ozone measurements during the winter of 1994–95 indicate that for middle and high latitudes in the Northern Hemisphere, ozone values were 10–20% lower than typical values observed during these months in 1979 and the early 1980s. Over some high latitude regions such as Siberia, total ozone had decreased by up to 35% from 1979 values. Over Northern Hemisphere mid-latitudes, total ozone decreased at the rate of about 4% per decade.
>
> *Northern Hemisphere Winter Summary*, April 1995, NOAA, Washington, DC.

The reason for stratospheric ozone depletion is largely explained by the photochemical decomposition of chlorofluorocarbons (CFCs) at low pressures and temperatures, although volcanic eruptions into the high atmosphere of sulphur and water vapour are contributing factors.

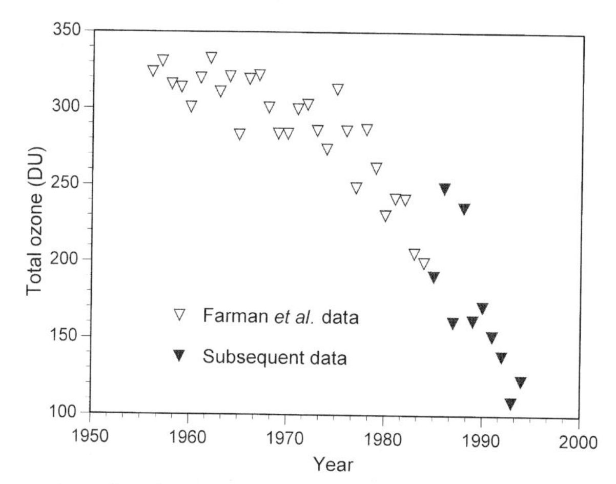


Fig. 5. October mean values of total ozone over Halley Bay Station, Antarctica. 1956–1984 (Farman *et al.*, 1985; Jones and Shanklin, 1995). DU stands for Dobson unit, a measure of the amount of ozone in the stratosphere. (one DU = 0.001 cm of ozone at standard temperature and pressure.) The empty triangles are from the original Farman *et al.* paper; the filled triangles are subsequent data.

There are two reasons to be concerned about stratospheric ozone depletion. In the first place, ozone is a shield that protects people and the biosphere from damaging ultraviolet B (UV-B) radiation from the sun. There is already some evidence that UV-B radiation is increasing, and that some biological material is being damaged (lichen in Greenland (Johnsen and Heide-Jørgensen, 1993), corals in the Caribbean (Gleason and Wellington, 1993) and phytoplankton off the coast of Antarctica). Daily weather forecasts in Australia and Canada now include predictions of UV-B intensity in relative terms.

Secondly, ozone and CFCs are 'greenhouse' gases, and thus are important in the context of global climate warming (see below).

That the Antarctic stratospheric ozone hole illustrated in Figure 5 is not a local phenomenon is demonstrated in Figure 6, which shows that the size of the 'hole' is continuing to increase.

### *Increasing Concentrations of the Atmospheric Greenhouse Gases*

The greenhouse gases include water vapour, carbon dioxide, methane, oxides of nitrogen, ozone and CFCs. These gases are almost transparent to downward radiation from the sun but they partially block upward radiation from the ground and clouds. This keeps global mean temperatures about 33 deg C warmer than they would otherwise be. In pre-industrial times, the greenhouse gases were more or less at equilibrium levels, except for long-term swings on time scales of millennia.

Atmospheric carbon dioxide concentrations began to rise in the 18th century (see Figure 7). This figure is a composite of direct measurements of atmospheric concentrations of carbon dioxide made on the top of Mauna Loa, Hawaii beginning in the 1950s, and indirect measurements from ice cores in Antarctica. Mauna Loa is above the

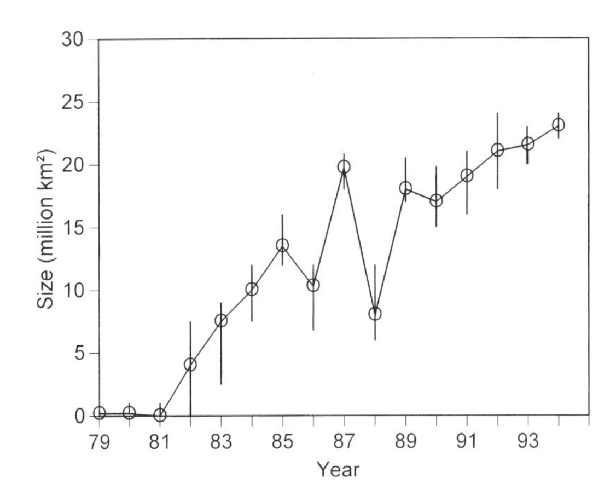


Fig. 6. Size of the stratospheric Antarctic ozone hole (Kerr, 1994, based on GSFC/NASA data).

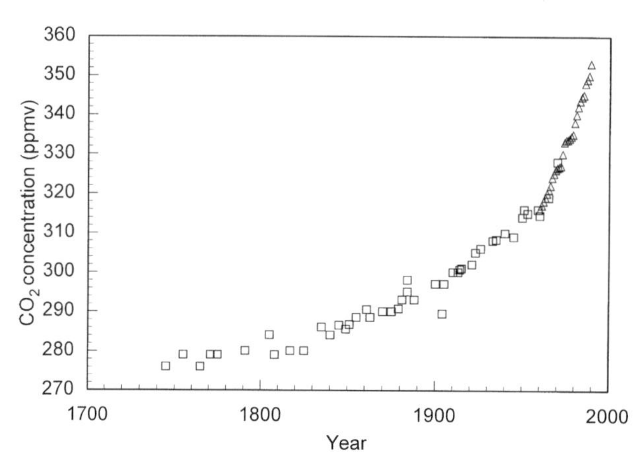


Fig. 7. Atmospheric $CO_2$ concentration, as indicated by measurements of air trapped in ice from Siple Station Antarctica (squares) and by direct atmospheric measurements from Mauna Loa, Hawaii (triangles) (Boden *et al.*, 1993)

Trade Wind inversion most of the time, and Antarctica is a long way from industrial sources so measurements there represent global trends. When the increasing atmospheric loading is compared with annual emissions of carbon dioxide from fossil fuel combustion, roughly half of the annual emissions stay in the atmosphere; the remainder is taken up by the land biosphere and the oceans.

Beginning in the 1970s, baseline measurements of some of the other greenhouse gases have been made, not only at Mauna Loa but also at other remote sites such as Point Barrow, Alaska and Cape Grim in Tasmania. These data sets all reveal upward trends that result from increasing emissions. In fact, current atmospheric levels of carbon dioxide, methane and nitrogen dioxide are unprecedented in the last 200 000 years.

Although higher concentrations of carbon dioxide should stimulate biological growth, and should cause the global mean surface temperature to rise, there is as yet no statistically significant evidence for global trends. Nevertheless, the trends in atmospheric concentrations of the greenhouse gases indicate that the world's climate may someday warm. Examining temperature trends over the last century, there is a hint that the Earth's surface temperature has risen over the last 20 years, but the amount is less than half a degree and is statistically not significant. Sea ice cover in the Arctic from 1961 to 1991 shows a decline in ice amounts in spring and summer, and the world's glaciers seem to be in retreat. According to the latest IPCC Summary for Policymakers (IPCC, 1995), "the balance of evidence suggests that there is (already) a discernable human influence on global climate".

## *Problems in the World's Forests*

Tropical deforestation losses between 1980 and 1990 are given in Table 1 for 87 countries in Africa, Latin America/Caribbean and Asia. The percentage losses may seem small (0.8 to 1.2%) but the total area lost for these 87 countries is 16.9 million hectares.

Table 1. Provisional FAO estimates of forest cover and deforestation for 87 countries in the tropical regions, as of 15 October 1991

| Continent | Number of countries studied | Forest area 1980 | Forest area 1990 | Annual deforestation 1981–90 | Rate of change 1981–90 (% year) |
|---|---|---|---|---|---|
| | | (millions of hectares) | | | |
| Africa | 40 | 650 | 600 | 5.0 | –0.8 |
| Latin America and Caribbean | 32 | 923 | 840 | 8.3 | –0.9 |
| Asia | 15 | 321 | 275 | 3.6 | –1.2 |
| Total | 87 | 1894 | 1715 | 16.9 | |

Notes: Countries include almost all of the moist forest zone, along with some dry areas. Figures are indicative, and should not be taken as regional averages.
Figures may not tally due to rounding.
Source: Dembner (1991) *Anasylva* **164**, 40–44.

In middle and high latitudes, deforestation is mostly a local problem, with afforestation taking place in parts of western Europe and eastern North America. In Scandinavia and the Laurentian shield of North America, high emissions of oxides of sulfur and nitrogen from industrial and automotive sources have led to the acidification of lakes and rivers, and some forest dieback (through mobilization of heavy metals as the soils become more acidic). Forests are also under stress from outbreaks of pests such as gypsy moths and spruce budworms. Spraying the pests has provided short-term relief but has interrupted long-term cycles in pest populations, worsening the problem.

Some of the consequences of forest destruction include effects on the hydrological cycle (e.g., the spring runoff is greatly intensified in areas having a winter snow pack), soil erosion, and loss of genetic resources.

### *Desertification*

Trends in desertification are difficult to assess. Using satellite imagery, Tucker *et al.* (1991) found that the Sahara expanded in the early 1980s but then contracted in response to higher rainfall. In China, recent desertification rates have been 0.6–1.6% per year, and 1.5–5.4% in the former USSR. This is leading to a loss of arable land and a reduction in food-producing capacity, particularly in marginal areas.

There is of course a relation between population growth and desertification. Figure 8, for example, shows the situation in Egypt. As the population increased from 10 million at the beginning of the century to more than 50 million at the present time, the *per-capita* arable land declined (the total amount of arable land increased slightly but this trend 'is highly unlikely to be maintained in the future' (Biswas, 1993)).

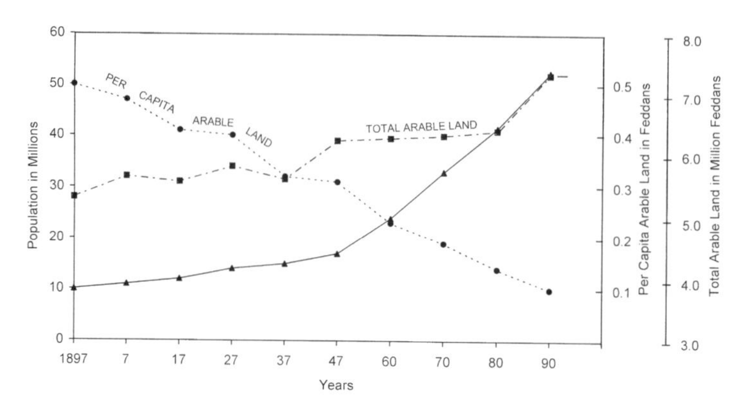


Fig. 8. Population (solid line) and arable land (total – dot/dashes; *per capita* – dashed line) in Egypt, 1897–1990 (Biswas, 1993). (1 feddan = 0.42 ha)

### *Frequencies of Natural Disasters*

Natural disasters include floods, droughts, tornados and tropical storms, heat waves and cold waves, earthquakes, volcanic eruptions, forest fires, avalanches and landslides, and insect infestations. Since 1963 when data were first collected systematically, the number of natural disasters around the world has continued to increase, and the impact on stricken regions has continued to worsen, especially in developing countries. In fact, the number of persons affected has been increasing at the rate of 6% per year in the last two decades (IDNDR Japan, 1994).

This is the United Nations International Decade on Natural Disaster Reduction, and efforts are being made worldwide to reduce the impact of natural disasters on society.

## Global Change in the 21st Century

Some examples of global trends that have already begun were given in the previous section. Evidence that these trends may become unprecedented in the 21st century comes from large computer simulation models. Here it should be emphasized that these models are particularly useful in providing answers to policy 'what-if' questions, such as 'How much climate warming would be expected *if* greenhouse gas concentrations doubled over their pre-industrial levels?'

Some people criticize simulation models, citing, for example, the Limits-to-Growth predictions of the 1970s. But these were scenarios, answers to 'what-if' questions, not predictions. Their real merit was in their effect on public attitudes, beginning to change the consumer society into the conserver society. Other examples of simulation models that had a major effect on public policy include:

- The nuclear winter scenarios: realization that nuclear war could have serious impacts on non-combatant countries in the Southern Hemisphere was a major factor in the outcome of the United Nations debates on nuclear test ban treaties.
- The greenhouse gas climate warming scenarios: it is perhaps surprising that First Ministers of many nations began to take climate warming seriously in the late 1980s, based almost entirely on simulation model predictions rather than on confirmed temperature rises, although there was firm evidence of strong increases in the atmospheric concentrations of the greenhouse gases.

One of the reasons that these simulation models had such a major impact on public policy is that the senior scientists involved (Dennis Meadows, Sir Frederick Warner, Bert Bolin, Steve Schneider and many others) were literate and persuasive.

Some illustrative examples of simulation model outputs are as follows.

### *Loss of Biodiversity*

Biodiversity has been on national and global political agendas for more than 20 years – for ethical reasons and because some species may some day be found to be valuable as medicines. In the last several years, a new and perhaps more compelling reason for preserving biodiversity has emerged – to ensure ecosystem functioning, particularly in a rapidly changing environment. Without a range of species, an ecosystem may have trouble surviving in a warmer and drier climate, for example. SCOPE (1996) and UNEP (1995) have recently completed extensive international literature reviews, which are recommended reading.

Rates of species extinction are difficult to estimate but there is no doubt that with increasing habitat destruction in all parts of the world, species diversity is bound to decline. In fact, a simulation model developed by ecologists at the University of Minnesota and Oxford University predicts that species extinctions may occur generations after landscape fragmentation, which means that there is a future ecological cost to current habitat destruction (Tilman *et al.*, 1994). A widespread consensus exists that species extinctions will increase in the 21st century, exacerbated by climate warming and other forms of global change. Biodiversity loss is of course an irreversible process, which adds to the urgency of the issue.

### *Climate Change*

Most atmospheric modellers believe that simply to stabilize atmospheric greenhouse gases at levels double those of their pre-industrial values, the global mean temperature will likely rise about 2–3 deg C, with warming greatest near the poles and lowest in the tropics. Of course, this may induce changes in cloudiness, precipitation and evaporation, which are difficult to estimate, as indeed are local changes, for given changes in global averages. These questions will be discussed in Chapter 4.

*Sea Level Rise*

Climate warming would cause the oceans to rise, and this would seriously impact on low-lying islands and coastal zones. Already, several international bodies are examining the severity of this threat, and the effectiveness of various proposed engineering solutions. At the same time, water levels in inland lakes and rivers may fall due to increased regional evaporation. The economic consequences would be substantial.

*Increasing Ozone Concentrations Near the Earth's Surface*

Ozone concentrations in the lower atmosphere have been increasing in the last two decades in contrast to stratospheric trends. Ozone is created near the ground through the reaction of oxides of nitrogen (from automobiles and agricultural activities) with volatile organic compounds (from various industrial processes and from wetlands) in the presence of sunlight. Ground-level ozone in sufficiently high concentrations is harmful to agricultural crops, forests and human health. In many parts of the world, ozone concentrations already exceed WHO standards from time to time – and, as pointed out by Chamiedes *et al.* (1994), the world's main food production areas (Europe, North America, eastern and southeastern Asia) are also the areas where surface ozone concentrations are increasing the fastest. Chamiedes *et al.* (1994) use a simulation model to project that, by the year 2025, 10–35% of the world's grain production will be at risk.

*Aggravating Factors*

Upward trends in human populations, particularly near coastal areas, in the consumption of natural resources and energy, and in the production of wastes, can only lead to serious effects on the biosphere and on society in the 21st century. In China, for example, primary energy demand is expected to more than double in the next quarter century, by which time it will exceed that of Western Europe (Searjeant, 1995). In the same period, the Chinese governent is planning to build 35 000 km of motorways; car ownership is presently increasing at 15% per year (Prynn, 1995). Here it should be emphasized that, in order to arrest rapid population increases in some of the Third World countries, industrial development would probably have to take place, increasing consumption of natural resources and waste production.

## What are the Consequent Critical Issues for Science and for Society?

In spite of the uncertainty, unprecedented change is definitely a possibility in the 21st century, and the consequences could be enormous. Global change must therefore be taken seriously. Principle 15 of the 1992 Rio Declaration states that:

> Where there are threats of serious or irreversible damage, lack of full scientific certainty shall not be used as a reason for postponing cost-effective measures to prevent environmental degradation.

'Global change' is a major issue for both science and society. The Brundtland Report (see p. 100) has brought an awareness of society's intergenerational responsibility to protect and indeed to enhance the world's store of natural resource (the so-called natural capital) for future generations. This is difficult enough in a world that is changing only slowly. In a rapidly changing environment containing uncertainties and surprises, the stakes are very high.

From an interview with Professor Bert Bolin published in the Dutch journal *Change*, **18**, pp. 4–7, February 1994:

*Question*: Not all scientists have the same opinion on the severity of climate change. Natural scientists and social scientists look at it in a different way and scientists who study the effects have a different view than the ones who study the global system. Would it be possible for scientists to come to a more unequivocal opinion? This would maybe help politicians to make decisions.

*Professor Bolin*: ... As far as the natural scientists are concerned, there is not much disagreement among them. I would like to hear more than one name of a scientist who has done work in this field – work which has been published in the scientific literature – who doesn't share the view of the IPCC *(that if we go on emitting greenhouse gases so that an equivalent doubling of carbon dioxide occurs, we would expect an increase of temperature for the earth as a whole by somewhere between 1.5 and 4.5 deg C).*

The critical issues for science and society are:

(1) The need for improved understanding of the Earth–Ocean–Atmosphere system, and of the demographic, socieconomic and cultural factors that contribute to global change. Some of the current international research programmes dealing with these questions are described in Chapters 2–4.

(2) The development of policies that will contribute to slowing down the changes, or to assist in human adaptation to these changes. Because the future is uncertain,

these policies must be developed in the framework of risk assessment and risk management.

(3) Promotion of the field of environmental engineering in the search for practical solutions to local as well as large-scale problems in both the public and private sectors.

(4) Development of long-term public education programmes that will lead to acceptance of tough political actions on a world-wide basis.

Global change issues include understanding and predicting the causes and impacts of, and potential responses to: long-term climate change and greenhouse warming; changes in atmospheric ozone and ultraviolet (UV) radiation; and natural climate fluctuations over seasonal to interannual time periods. Other related global issues include desertification, deforestation, land use management, and preservation of ecosystems and biodiversity.

From *Our Changing Planet, the FY95 U.S. Global Change Research program*, p. 2, a report of the United States National Science and Technology Council, Washington, 1994

Table 2. Increasing frequencies of the use of key words in selected scientific journals

| Keyword | 1983 | 1988 | 1993 |
|---|---|---|---|
| Global change* | 0 | 29 | 174 |
| Climate change* | 0 | 54 | 114 |
| Global warming* | 0 | 9 | 77 |
| Biodiversity** | 0 | 4 | 134 |

* From GEOREF, the CD ROM version of the American Geological Institute's geoscience database.
† From the Life Sciences Collection of a CD ROM containing records from 21 abstract journals published by the Cambridge Scientific Abstracts.

## Concluding Remarks

The main points to be emphasized in this brief overview are that:

(1) There are already clear signals that society is disturbing the Earth System's ability to support life.

(2) Since we do not know enough of how the Earth System works, a concerted international interdisciplinary research effort is crucial, supported by global monitoring, modelling and other appropriate tools. The aim is to provide a basis for designing strategies for prevention, mitigation, adaptation and possibly exploitation of global changes.

(3) The evidence is persuasive that business, industry and governments should adopt the precautionary principle, and should routinely include 'global change' in their long-range planning activities.

## Selected References

Ayibotele, N. B. and Falkenmark, M. (1992) Freshwater resources, Chapter 10 in *An Agenda of Science for Environment and Development into the 21st Century*, J. C. I. Dooge, G. T. Goodman, J. W. M. la Rivière, J. Marton-Lefèvre, T. O'Riordan and F. Praderie, eds. Cambridge University Press, Cambridge, U.K., pp. 187–203.

Baxter, D. and Laglagaron, D. (1993) *Livable Region Strategic Plan Growth Management: A Compact Metropolitan Vancouver Region Option*, Greater Vancouver Regional District, Burnaby, BC, 46 pp.

Biswas, A. K. (1993) Land resources for sustainable agricultural development in Egypt. *Ambio*, **22**, 556–560.

Boden, T. A., Kaiser, D. P., Sepanski, R. J. and Stoss, F. W. (eds.) *Trends '93*, ONRL/CDIAC-65, Carbon Dioxide Information Analysis Center, Oak Ridge Nat. Lab., Oak Ridge, Tenn., USA.

Bojkov, R. D. (1992) Changes in polar ozone. *WMO Bulletin*, **41**, 171–180.

Boutron, C. F., Gorlach, U., Candelone, J.-P., Bolshov, M. A. and Delmas, R. J. (1991) Decrease in anthropogenic lead, cadmium and zinc in Greenland snows since the late 1960s. *Nature*, **353**, 153–156.

Chamiedes, W. L., Kasibhatla, P. S., Yienger, J. and Levy II, H. (1994) Growth of continental-scale metro-agroplexes, regional ozone pollution, and world food production. *Science*, **264**, 74–77.

Dynesius, M. and Nilsson, C. (1994) Fragmentation and flow regulation of river systems in the northern third of the world. *Science*, **266**, 753–762.

Farman, J. C., Anderson, J. G., Toohey, D. W., Fahey, D. W., Kawa, S. R., Jones, R. L., McKenna, D. S. and Poole, L. R. (1985) Large losses of total ozone in Antarctica reveal seasonal $ClO_x/NO_x$ interaction. *Nature*, **315**, 207–210.

Gleason, D. F. and Wellington, G. M. (1993) Ultraviolet radiation and coral bleaching. *Nature*, **365**, 836–838.

Gunn, J. M. and Keller, W. (1990) Biological recovery of an acid lake after reductions in industrial emissions of sulphur. *Nature*, **345**, 431–433.

Hong, S., Candelone, J.-P., Patterson, C. C. and Boutron, C. F. (1994) Greenland ice evidence of hemispheric lead pollution two millennia ago by Greek and Roman civilizations. *Science*, **265**, 1841–1843.

Hong, S., Candelone, J.-P., Patterson, C. C. and Boutron, C. F. (1996) History of ancient copper smelting pollution during Roman and medieval times recorded in Greenland ice, *Science*, **272**, 246–249.

Hilborn, J. and Still, M. (1990) *Canadian Perspectives on Air Pollution*. SOE Report No. 90–1. Environment Canada, Ottawa, Canada, 81 pp.

IDNDR Japan (1994) *Disasters Around the World – A Global and Regional View*. Info. paper No. 4, World Conference on Natural Disaster Reduction, Tokyo, 23–27 May 1994, 87 pp.

IPCC (1995) Working Group I Summary for Policymakers, IPCC, WMO, Geneva, Switzerland, 7 pp.

IUCN/UNEP/WWF (1991) *Caring for the Earth: a Strategy for Sustainable Living*. IUCN, Gland, Switzerland.

Johnsen, I. and Heide-Jørgensen, H. S. (1993) *Impact of increased UV-B radiation on the lichen Cladonia mitis from Northern Greenland and Denmark*. Interim report of a research project, Botanical Institute, Department of Plant Ecology, University of Copenhagen, Denmark, 15 pp.

Jones, A. E. and Shanklin, J. D. (1995) Continued decline of total ozone over Halley, Antarctica since 1985. *Nature*, **376**, 409–411.

Kerr, R. A. (1994) Antarctic ozone hole fails to recover. *Science*, **266**, 217.

MacKenzie, J. (1991) *Global Trends in Passenger Cars*. Climate Institute, Washington, DC.

Postel, S. L., Daily, G. C. and Erhlich, P. R. (1996) Human appropriation of renewable fresh water. *Science*, **271**, 785–758.

Prynn, J. (1995) UK in slow lane to build Chinese mega motorway, *The Times*, Apr. 3 (Business Section).

SCOPE (1996) *Ecosystem Function of Biodiversity* (H. A. Mooney, ed.) John Wiley, Chichester, U.K. (In press).

Seargeant, G. (1995) Keep the lights burning in Walmington-on-sea, *The Times*, Apr. 10, p. 42.

Tilman, D., May, R. M., Lehman, C. L. and Nowak, M. A. (1994) Habitat destruction and the extinction debt. *Nature*, **371**, 65–66.

Tucker, C. J., Dregne, H. E. and Newcomb, W. W. (1991) Expansion and contraction of the Sahara desert from 1980 to 1990. *Science*, **253**, 299–301.

UNEP (1992) *The World Environment 1972–1992*. Chapman & Hall, London, 884 pp.

UNEP (1995) *The Global Biodiversity Assessment*. UNEP, Nairobi, Kenya and Cambridge University Press, Cambridge, UK, 1140 pp.

Walsh, M. (1990) Global trends in motor vehicle use and emissions. *Annual Review of Energy*, **15**, 217–243.

# 2. The Main International Environmental Research Programmes

J. W. M. LA RIVIÈRE
*IHE, Delft, The Netherlands*

## Introduction

The previous chapter has demonstrated that global change has become an important scientific and political issue that will not go away. Rather, it will unfold during the next decades not only by creeping alterations but also by sudden surprises.

The gradual realization that global change is already taking place is, of course, the result of scientific research and observation, through which the first signals were detected and interpreted. Now that decision-makers have recognized its importance, a dialogue between science and policy-making has begun and – understandably – the first question asked of science is 'What exactly is going to happen and when?' Decision-makers demand forecasts and, euphemistically, 'removal of uncertainties'. Addressing this question requires a comprehensive understanding of how the Earth System operates, its life-support functions, and how it responds to disturbances. Never before had this question been asked in its totality, science having addressed the Earth System in a disciplinary fragmentary fashion. Fulfilling the new mandate thus required a novel global and interdisciplinary effort specifically designed for the purpose; also it had to be well connected to the policy-makers to provide them with the best available scientific knowledge as a basis for strategy making.

Decision-makers while insisting on quick answers nevertheless realized full well that these could not be produced overnight, and that the answers would be bound to contain some uncertainty and surprises. Decision-makers therefore adopted the 'precautionary principle' (if the consequences are completely unacceptable, then the threat should be examined within the framework of risk assessment and risk management), which would, hopefully, slow down some aspects of global change and thus buy time for doing the urgently required research.

During the past decade or two, new scientific efforts took shape that made use of what was already in place and added what was lacking. In this process, new programmes were

*R. E. Munn, J. W. M. la Rivière and N. van Lookeren Campagne (eds), Policy Making in an Era of Global Environmental Change, 17–22.* 

shaped at the international, regional (e.g., European), and national levels, enmeshed with existing programmes, each with its own history, constituency and – sometimes redirected – mandate. The resulting complex ensemble at first looks like an 'alphabet soup' of acronyms (see Appendix 2), an aggregate of elements of different sizes and character. However, during this evolution – or shakedown period – the natural forces of shortages of qualified scientists and available resources have streamlined the sum total of the research efforts into a system in which division of tasks, interconnections and synergistic cooperation prevail.

As global change will lead to 'winners and losers' among nations, the global response strategies, e.g., in the form of conventions, are necessarily the result of intensive political negotiations in which national interests play a significant role. It is, therefore, of vital importance that undisputed objective scientific information be the basis of such negotiations. This is one of the reasons why governments and especially the UN rely to a large extent on the work of non-governmental, non-advocacy scientific organizations, of which ICSU is most prominent.

The objective of this monograph is not to describe comprehensively the entire field of global change research activities in all countries. Instead we concentrate on two major international research programmes (IGBP and WCRP) now in operation and recognized by the United Nations General Assembly. Then we present a brief overview of some other international programmes. More information about IGBP and WCRP is given in Chapters 3 and 4, written by the Chairman and past-chairman of their respective scientific steering committees.

Addresses of the main bodies active in the field of global environmental change are given in Appendix I.

## The IGBP and WCRP: Structure and Manner of Operation

The scientific issues addressed by IGBP and WCRP will be explained in Chapters 3 and 4. Here follows some information about their structure and manner of operation.

Both programmes are led by a steering committee composed of high-level scientists. These are appointed for specific periods of time by the programme sponsors, who aim at balance with respect to nationality and disciplinary expertise as required for the committees' work. For IGBP, ICSU is the only sponsor. WCRP is jointly sponsored by WMO, ICSU and IOC of UNESCO. The sponsors also appoint a scientific director who heads the executive offices (secretariat) of the programmes, under the responsibility of the steering committees. For WCRP, the secretariat is housed in Geneva with WMO; IGBP has its offices in Stockholm at the Royal Swedish Academy of Sciences. Each has a small staff (less than 10).

The research agendas for the programmes are prepared by the steering committees, who identify priority issues and devise realistic ways of addressing them through a coordinating framework within which national research efforts can be synergistically deployed. Or phrased differently, scientists from different countries jointly prepare masterplans into which their countries find it to their advantage to dovetail their national research activities. This leads to a division of tasks and to economy in the use of manpower and equipment like satellites and research vessels.

Developing research agendas and coordinating frameworks is a large and complex operation requiring many consultations and workshops, and the preparation of many documents. In addition, the steering committees and secretariats have to monitor the implementation of research and the reporting of results which then have to be synthesized, interpreted and communicated to policy-makers. Moreover, in the light of preliminary results obtained, the agenda has to be continually adjusted, while also new issues may require attention. It is, therefore, not surprising that both programmes have, to some extent, compartmentalized their work in a number of separate projects as illustrated in the following chapters. In order to maintain coherence, special interactive mechanisms between the projects have been put in place.

In addition to the tasks mentioned above, there is need to maintain active contact with the sponsors, with consultative bodies, governmental and intergovernmental organizations, funding agencies, and other relevant international research programmes. Most importantly, IGBP and WCRP have to take part in discussions with space agencies and global observing systems in which they are invited to put forward their specific data needs from, e.g., future satellite missions.

The scientific secretariats have begun to function as centres of networks, both within their own programmes and the scientific community at large, greatly assisted by the networks of their own sponsors, including that of ICSU with its 23 Unions, 94 national Academies and 30 scientific associates.

Besides FAX, telephone and electronic mail, newsletters, electronic bulletin boards and face-to-face meetings keep the networks alive, functioning and responsive.

It is especially important that WCRP works closely with IGBP because they are jointly managing all aspects of climate change research, an important segment of global change research; WCRP deals with the physical aspects while IGBP includes the chemical and biological aspects. Effective coordination is achieved through substantive *ex-officio* seats on the two steering committees and intensive communication between offices.

## Some Other Relevant Research Programmes

There are, of course, other relevant research programmes relating to global environmental change. A few of the important ones will be highlighted here:

### *Diversitas (ICSU–SCOPE–UNESCO/MAB)*

The international research programme *Diversitas* focuses on biodiversity. Not only is the rapid and irreversible loss of genetic resources an ethical and an economic issue, but also such losses may remove 'keystone' organisms which affect the functioning of ecosystems. Since humankind literally does not know what it is doing by destroying genetic resources, research is badly needed. The importance of the issue has been recognized by the recent ratification of the Biodiversity Convention, drawn up and signed in Rio in 1992.

*International Human Dimensions of Global Environmental Change (IHDP) (ISSC – International Social Science Council and ICSU)*

The international research programme on the Human Dimensions of Global Environmental Change aims to respond to the widely heard general statement that since humankind is both the cause and the victim of global change, study of the socio-economic driving forces of global change and of its impacts on society are urgently needed. Since the – relatively small and very diverse – social science community is not strongly organized internationally and has had no experience with large research programmes of the type that seems to be required, the Human Dimensions programme has wisely taken a cautious approach in its initial agenda-setting phase. The first projects likely to become operational in the near future will be undertaken in collaboration with IGBP and will deal with two IGBP projects: Land Use and Land Cover (LUCC) and Land-Ocean Interaction in the Coastal Zone (LOICZ). Furthermore, it is clear that societal impacts and adaptation scenarios offer a vast field of study for both international and national social science research, and for collaboration with the natural sciences (la Rivière, 1991). The Human Dimensions programme closely collaborates with IGBP and WCRP, e.g., in the development of the Global System for Analysis, Research and Training (START) for global change.

*Global observation systems (GCOS/GOOS/GTOS)*

Existing environmental observation systems were designed for specific purposes, but in the early 1980s it became clear that their sum total would not be capable of monitoring the Earth System as a whole to meet the need of global change research, to upgrade environmental forecasts and to detect global change against the 'noise' of natural variability. For instance, observing systems for *weather* prediction obviously require extension to meet the needs of *climate* monitoring and prediction. Hence, three interconnected global observation systems are now being designed jointly by ICSU and relevant UN organizations. These are:

GCOS: Global Climate Observing System
GOOS: Global Ocean Observing System
GTOS: Global Terrestrial Observing System.

Together they will keep the Earth System in its totality under observation, making full use of existing systems and supplementing them as needed.

*SCOPE (Scientific Committee on Problems of the Environment)*

Special mention should be made of the Scientific Committee on Problems of the Environment (SCOPE), an ICSU body that by itself does not undertake research, but has the role of analysing and assessing global environmental problems on the basis of scientific information drawn from all the disciplines involved. This is an urgent task because of the enormous rate at which the scientific literature is expanding – more than 2 million papers are published every year in science and medicine.

Through a series of workshops, international interdisciplinary teams of scientists synthesize the available information, and after critical analysis, identify knowledge gaps and propose research priorities. Since its creation in 1969, SCOPE has published more than 50 monographs, many of which in retrospect have proved to be pioneering efforts. Examples are:

*SCOPE 2, 1972: Man-made lakes as modified ecosystems.* This is a dispassionate inventory of the pros and cons of constructing large reservoirs. The subject is of great interest since, for example, energy production without using fossil fuels is important with respect to global warming.

*SCOPE 5, 1979: Environmental impact assessment: principles and procedures.* This 'best-seller' helped to develop EIA as an instrument of environmental policy in many countries.

*SCOPE 21: The major biogeochemical cycles and their interactions.* This monograph drew the attention of both scientists and policy analysts to the fact that many environmental issues (e.g., climate warming, stratospheric ozone depletion, acid rain) were interrelated through the global biogeochemical cycles. The monograph laid the foundation for SCOPE 29.

*SCOPE 29, 1986: The greenhouse effect, climatic change, and ecosystems.* The results of this study led governments to place Climate Change on the international political agenda.

*SCOPE 28, 1985, 1986: Environmental consequences of nuclear war.* This study showed that nuclear explosions in the Northern Hemisphere would have serious consequences for countries in the Southern as well as the Northern Hemispheres, through the 'nuclear winter' phenomenon. There are indications that these two volumes influenced nuclear disarmament discussions in the UN.

*SCOPE 40, 1989: Methods for assessing and reducing injury from chemical accidents.* This report is based upon case studies of major accidents in the chemical industry, such as the one that occurred at Bhopal.

*SCOPE 50, 1993: Radio-ecology after Chernobyl.* This study has considerably increased our understanding of biogeochemical pathways of artificially-produced radionuclides. Beside the Chernobyl accident, release of radionuclides from other sources such as from nuclear weapon tests and accidents at Windscale, Kyshtym and Three Mile Island was studied.

Other examples of SCOPE reports deal with the effects of toxic chemicals and UV-B radiation on ecosystems and humans. Such studies are often done in collaboration with WHO and UNEP, and have helped to make ecotoxicology a recognized area of science.

### *Other*

Many other ICSU bodies and some other inter-governmental and NGO bodies have research activities wholly or partially devoted to aspects of global change. In fact, the total effort on global change research is so large that it has been recognized by the OECD as belonging to the category of megascience.

## Selected References

IGBP (1994) IGBP in Action: Work Plan 1994–1998. Report No. 28, IGBP Stockholm, Sweden, 151 pp.

la Rivière, J. W. M. (1991) Cooperation between natural and social scientists in global change research: imperatives, realities, opportunities. *Int. Social Sci. J.*, **130**, 619–627.

See also Chapters 3 and 4 of this monograph.

# 3. Understanding the Earth System

P. WILLIAMSON and P. S. LISS
*IGBP Chairman's Office, School of Environmental Sciences, University of East Anglia, Norwich NR4 7TJ, UK*

## Introduction: Cycles and Systems

Viewed from a global perspective, biological processes use raw materials at a rate that greatly exceeds their supply from the planetary interior – a situation that can only be sustained through a complex series of chemical energy transfers, linked to the global movements of materials, from land to sea, within the oceans, and through the air. The variety and abundance of life at the Earth's surface would not exist without this recycling.

Consider oxygen. By weight, this is the most important single element within plants and animals (for example, around 65% of human body mass), and its ready availability is essential for the metabolic activities of all multicellular organisms. Whilst gaseous oxygen is plentiful in the atmosphere, with a total weight of around a thousand million million tonnes, that reservoir is subjected to a variety of removal and addition processes. Biological activity is particularly important in these exchanges, using oxygen, time and time again. In every breath we take, there are around $10^{19}$ atoms of oxygen; it is therefore a statistical near-certainty that at least one of these oxygen atoms has been previously breathed by Confucius – and another by Albert Einstein, or anyone else who lived more than a few decades ago (to allow for worldwide mixing within the atmosphere). Many more atoms of oxygen in each breath will once have been part of bacteria, beetles and birds, for example; or trees in the Amazon rain-forest, mosses in the Arctic tundra, or microscopic algae in the Pacific Ocean.

Free oxygen is not only used by life, but is itself a product of life. Any oxygen occurring in the Earth's primordial atmosphere would have been rapidly used up by mineral oxidations at the land surface – by the rusting reactions that change ferrous to ferric iron, and sulphides to sulphates. When samples of Moon rock, vacuum-sealed by the crew of Apollo 11, were exposed to air on Earth, they rapidly darkened and gave off a sulphurous smell that was likened to burning gunpowder. Our planet's 1500 million cubic kilometres of water, in oceans, lakes and rivers, provide the source of present-day atmospheric oxygen, released from its bond with hydrogen by biologically driven reactions. On average, each water molecule has a lifetime of around two million years before it is split by plant photosynthesis, on land or at sea, using sunlight as the energy source. Oxygen is liberated in that process, whilst the hydrogen is combined with carbon dioxide to form organic compounds.

*R. E. Munn, J. W. M. la Rivière and N. van Lookeren Campagne (eds), Policy Making in an Era of Global Environmental Change, 23–55.* 

The 'opposite' process of respiration, by plants, animals and microbes, enables the captured energy to be used for building cellular constituents, and for the wide range of other metabolic functions that characterize life. Animals use much of that energy for movement. Some is also used to generate electricity, for internal signalling and information management. The net effects of respiration are, eventually, the release of heat, the reunion of oxygen with hydrogen, and the return of water and carbon dioxide to the environment.

**Box 1. The natural economy**

Global trade, sustainable development, competition and recycling are nothing new: that has been the way of the natural world since life began on this planet. At least 25 chemical elements are required for biological processes, and for more than 3000 million years these basic commodities have been packaged, distributed, and traded worldwide within the natural economy. All parts of this interconnected system depend, directly or indirectly, on production companies that obtain their raw materials from air, rocks and water. The initial processing of these inanimate substances requires energy, derived from a nuclear fusion reactor, 150 million km away. Some of the resulting energy-rich organic compounds are retained by the producers, others are passed to traders, either for their direct use or for modification and upgrading. Processes for improving manufacturing techniques and functional design are built into the infrastructure of the natural economy, through both competition and partnerships. The overall effect is to maximize efficiency whilst favouring diversification, with the development of an extraordinarily wide range of products and packaging. After an operational period varying from a few hours to many years, the organic fabrications are broken down and their components are returned to the non-living world.

But the natural economy is now in trouble. The spectacular success of one high-growth company is unsustainable, being based on the depletion of capital assets that had, until recently, been held in common ownership. As a result, the market has been distorted: many other well-established companies, both producers and suppliers, have been driven out of business or are now trading at a loss, and the natural economy as a whole faces instability and future insolvency.

If the supply of solar energy to the Earth were to cease, plant life would rapidly come to an end, and the demise of animal life would soon follow. Microbes involved in organic decomposition and other chemical breakdown processes would undoubtedly persist; however, atmospheric oxygen would no longer be replenished. Mineral oxidations would then progressively use up the oxygen in the air, so that within a few thousand years, levels would fall from the present value (21%) to the trace amounts present in the pre-Cambrian atmosphere. In the unlikely event that the human species survived this science-fiction scenario, we would be faced with a planet nearly as inhospitable as Mars.

In addition to oxygen, hydrogen and carbon, the elements needed to sustain life include nitrogen, phosphorus, sulphur, calcium, sodium, potassium, iron and a wide range of trace substances. These are all involved in global cycles, with a series of exchanges and transformations between living and non-living states. Indeed, chemical activity at the Earth's surface is now dominated by such reactions – with its 'thin green film' exerting a major control on the fluxes of energy and materials at the land and sea surfaces, and on the composition of the oceans, the atmosphere, soil and sediment. Biological processes hasten rock weathering, through the acids produced in soil by respiration and decomposition; however, they slow the erosional transport of surface material from land to sea, through the binding together of rock particles, and the trapping of silt and sediment by the vegetation of river banks, flood plains, marshes and the coastal fringe.

Many different systems, operating at different scales, are involved in these natural cycles of elements. These have been separately investigated by a wide range of scientific disciplines: organic reactions at the molecular and cellular level have been studied by biochemists and microbiologists; transformations and transport within tissues, organs and organisms, by plant and animal physiologists; large-scale nutrient dynamics, energy flows and the interactions of species and communities, by agriculturalists and ecologists; and the physico-chemical framework for all these processes has been addressed by geologists, geographers, oceanographers, hydrologists and atmospheric scientists. Awareness of the continuum of these disciplinary approaches first developed in the geological community. Thus, an Austrian geomorphologist, Eduard Suess, introduced the term *biosphere* in 1875, whilst a Russian mineralogist, Vladimir Vernadsky, elaborated the concept in the 1920s. Vernadsky defined the biosphere as that part of the Earth where life exists: the envelope around the planet where there is water, an external supply of energy, and interfaces between the liquid, solid and gaseous states of matter. But another fifty years passed before there was wider interest in biogeochemical processes, and the global budgets of energy and materials that constitute the Earth System.

That interest has, of course, been greatly stimulated by public concern regarding pollution and such issues as the enhanced greenhouse effect and stratospheric ozone depletion. Human activities throughout the world disturb the status of natural cycles by accelerating some pathways and slowing others (Figure 1), altering the strength of sources and sinks of key elements. Society also adds to the environment, either accidentally or deliberately, a large number of novel, manufactured chemicals (mainly organic). Concern regarding pollution hazards initially focused on overtly toxic chemicals, such as pesticides and organic mercury compounds, and their relatively localized impacts on natural ecosystems and human health. However, it was later recognized that potentially more serious problems, with worldwide effects, can be caused by altering the levels of naturally-occurring and apparently harmless compounds (such as carbon dioxide), and by the release of synthetic chemicals that had been specially selected for their inertness and low toxicity (such as the chlorofluorocarbons, CFCs).

Political leaders, industrialists and ordinary citizens throughout the world are now increasingly aware that pollution and the careless exploitation of natural resources can affect our future comfort and economic prospects, as well as threatening the very existence of many millions of species with which we share the planet. Recognition of the

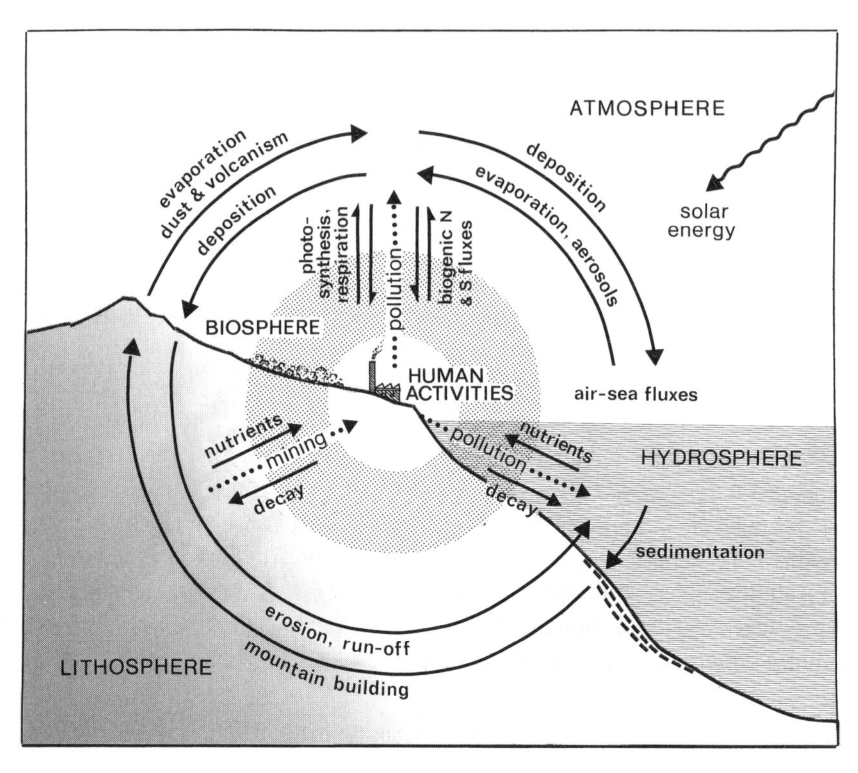


Fig. 1. Earth System cycles: the main mechanisms moving materials between land, air and oceans, and their principal interactions with biological processes and human activities. After S. D. Deevey, 1970 (*Scientific American*, **223**, 148–158).

value of environmental assets is relatively easy; taking effective action to halt or reverse damaging trends is much more difficult. That is particularly true for global change problems, requiring policy agreement at the international level, under conditions of considerable uncertainty. Those information gaps are not only of a basic, scientific nature, but also concern the long-term implications for energy consumption, land use, North–South relationships and inter-generational equity. Earth System research is directed at narrowing these key uncertainties that limit our understanding of global change. Whilst the emphasis is on the fundamental processes – how the system operates – such aspects must be considered in the context of socio-economic needs and policy concerns. It is unrealistic (and unnecessary) to aim for full predictability of such a complex meshing of natural and novel effects, inherent variability and human behaviour. Nevertheless, there is immense scope for replacing a state of ignorance, when all possibilities seem equally likely, with knowledge of risks, providing a much sounder basis for decision-making – in what will remain an uncertain world.

The investigation of living/non-living interactions at the planetary scale is a task that clearly benefits from an interdisciplinary and international approach. Furthermore, global integration of that effort requires scientific partnerships between industrialized and developing nations, to ensure full geographical coverage and data availability, and also to help in reaching agreement on the interpretation and assessment of scientific findings. The International Geosphere–Biosphere Programme (IGBP), established in 1986 by the International Council of Scientific Unions (ICSU), provides the research agenda and organizational framework to meet these challenges (see Box 2). IGBP effort is focused on a set of well-defined global change problems that have greatest worldwide significance for a future timescale of 10–100 years. It involves researches in all the disciplines identified above, working with system analysts, modellers, and experts in Earth observation and data management. IGBP also has close links with physical climatologists (through the World Climate Research Programme, WCRP), and with social scientists addressing global change issues (through the International Human Dimensions of Global Environmental Change Programme, IHDP). Although IGBP is non-governmental, it has been endorsed by the UN, and it is highly regarded by other international organizations, government agencies and industry. Funds for programme coordination are provided by those bodies, by science academies (through IGBP's national committee structure), and by ICSU, with the research itself being mostly funded at the national level.

**Box 2. The International Geosphere Biosphere Programme: a Study of Global Change (IGBP)**

The aim of IGBP is to describe and understand the interactive, physical, chemical and biological processes that regulate the total Earth System, the unique environment that it provides for life, the changes that are occurring in this system, and the manner in which they are influenced by human activities.

The planning and implementation of the programme is currently based on eight major research projects: Biospheric Aspects of the Hydrological Cycle (BAHC), Global Change and Terrestrial Ecosystems (GCTE), Global Ocean Ecosystem Dynamics (GLOBEC) International Global Atmospheric Chemistry project (IGAC), Joint Global Ocean Flux Study (JGOFS), Land–Ocean Interactions in the Coastal Zone (LOICZ), Land Use/ Cover Change (LUCC) and Past Global Changes (PAGES). These projects are linked by three framework activities: a Data and Information System (IGBP-DIS); a Task Force on Global Analysis, Interpretation and Modelling (GAIM); and the System for Analysis, Research and Training (START). Close partnerships exist with many other international programmes, including project co-sponsorship with the International Human Dimensions of Global Environmental Change Programme (for LUCC and START), the World Climate Research Programme (for START), the Scientific Committee on Oceanic Research (for JGOFS), and the International Association of Meteorology and Atmospheric Sciences (for IGAC).

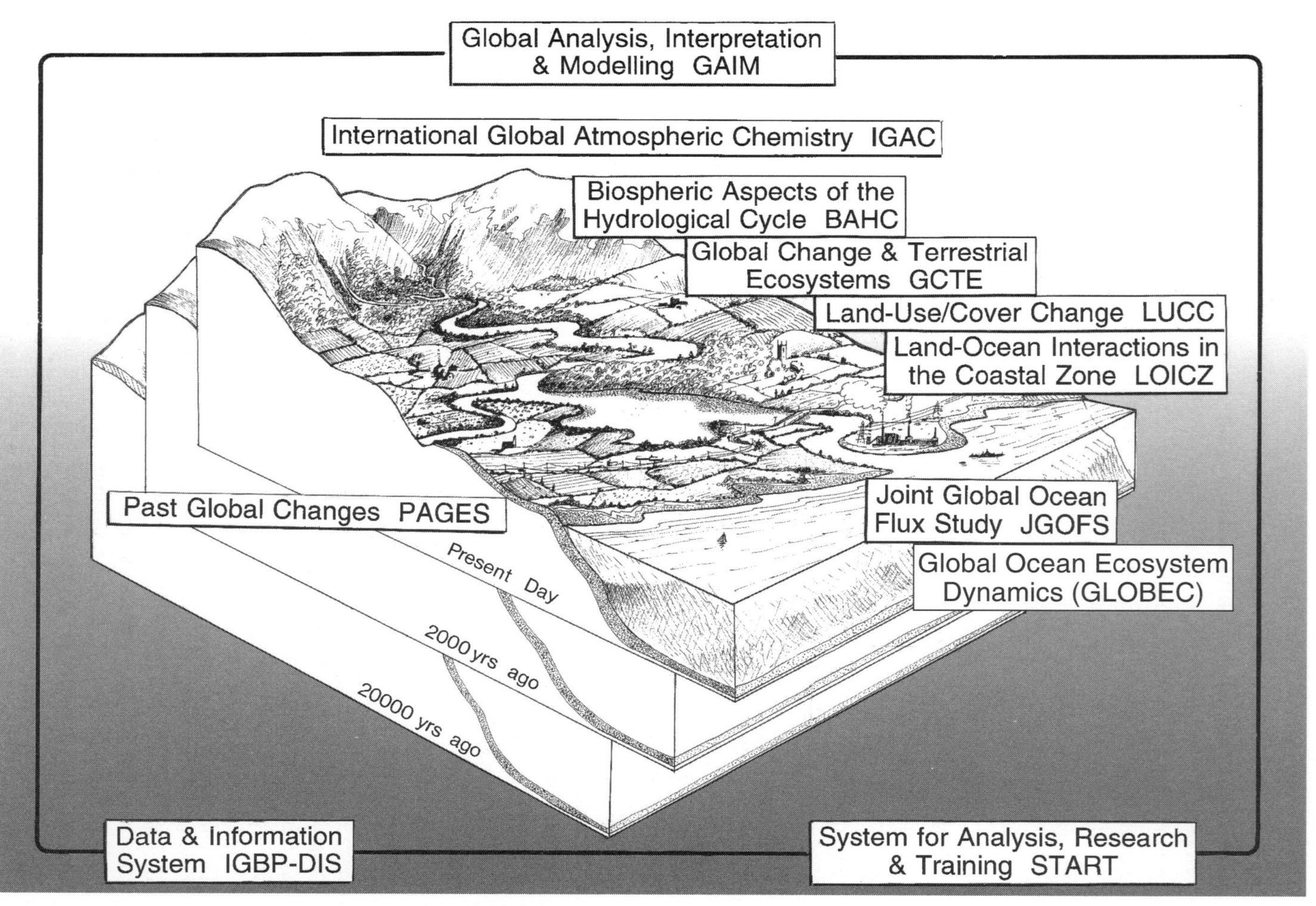


Fig. 2. Components of the International Geosphere–Biosphere Programme.

As an article in *The Economist* (5 November 1994, p. 119), concluded "Understanding the world is one thing; living in it is another". It is the human dimension that brings global change problems into focus – as recognized by ICSU when it agreed to share responsibility for the IHDP in 1996.

This chapter addresses some key features of the interplay of biospheric and geospheric processes, and the importance of those interactions for understanding global change. The physical climate system, considered in detail in the following chapter, cannot be excluded – since the processes that distribute energy and materials are fundamental to the operation of the Earth System, determining where life is possible and in what form. Thus the 'climatological' average conditions for temperature and rainfall strongly affect the geographical distributions of species, ecosystems and many human activities. Superimposed, there are multi-annual trends, extreme events and seasonal differences from year to year, all of which can affect the biodiversity and functional properties of natural ecosystems. Such variability also alters the productivity of farmland, fisheries and managed forests, and has other economically important consequences for freshwater resources, soil erosion, flood risks (particularly in coastal regions) and human health.

But Earth System science is not just a one-way study of climatic impacts. Biological changes at the land and sea surface affect other components of global systems, including both positive and negative feedbacks on physical processes. Whilst there is still debate regarding the magnitude of human influences on climate within the next 20–50 years, other far-reaching changes in global biogeochemistry have already occurred. These will intensify with the near-certain doubling of the human population by the middle of the 21st century. Even in developed countries, with stable populations, expectations of economic growth and ever-increasing living standards add to the pressures on environmental systems.

## 'Nothing is Permanent Except Change' (Heraclitus, c. 500 BC)

An individual's track record is widely used as a guide to how he or she might behave when provided with new challenges or opportunities. A similar approach provides revealing insights into the behaviour of the Earth System. In the absence of any opportunities for global-scale experiments (other than of an unintentional nature, without scientific controls), the different combinations of circumstances that have occurred in the past provide the most practical way to test our knowledge of unifying principles. Such a mix of empirical and theoretical understanding is essential for developing realistic 'scenarios' of future conditions, providing the information needed by policy-makers at the national and international level.

Both change and constancy feature strongly in the Earth's *curriculum vitae*. In this context, climatic behaviour is often used as a convenient shorthand for the wider environmental changes that have occurred during the recent and more distant past. For all that time, life – requiring free water – has continued. We can therefore assume that, despite their climatic buffeting, the oceans have never boiled dry, nor completely frozen. Since life has not merely survived but (judging from the fossil record) has generally flourished over the past 3000 million years, large areas of land and ocean must have experienced a considerably narrower, and more comfortable, range of conditions: Earth has fortunately avoided the fate of either a runaway greenhouse effect (as on Venus, with surface temperatures around 480°C), or perpetual frost (as on Mars, at –50°C), or large day–night temperature fluctuations, following the loss of atmosphere (as on the Moon, with a diurnal temperature range of up to 300°C).

The Gaia hypothesis (see Box 3) argues that life itself is intimately involved in regulating the planetary thermostat. One possible mechanism might involve changes in cloud cover; in particular, its potential control by biologically produced sulphur emissions from the oceans. Several species of marine phytoplankton produce sulphur-containing compounds as osmotic regulators, and, when they die or are eaten, the gas dimethylsulphide (DMS) is released. DMS forms acid aerosols in the atmosphere, driving many important chemical reactions and providing condensation nuclei that play a major role in cloud formation. The global total of oceanic sulphur emissions is estimated to be 15–20 million tonnes per year – considerably more than the amount released by volcanic activity

**Box 3. Gaia: science or speculation?**

A relatively narrow range of climatic conditions has been experienced on Earth since life first appeared. The survival of life defines that range, since biochemical constraints set body temperature tolerances at between 0°C–40°C for most organisms. The Gaia hypothesis, developed by James Lovelock and Lynn Margulis, proposes that global thermo-regulation processes, involving plants, animals and microbes, have been responsible for keeping the climate within tolerable limits. On that basis, the behaviour of the Earth System is itself capable of evolution, favouring the conditions that are best suited for building increasingly complex links between animate and inanimate processes.

These ideas have attracted much interest – but also criticism. If conditions on Earth had become much cooler or warmer, or if the chemistry of the atmosphere had been very different, would not life have developed in different ways, with appropriate adaptations to the new constraints? That question is probably unanswerable. Nevertheless, other aspects of the Gaia hypothesis are testable, and have provided important scientific insights. For example, it is now generally agreed that life is absent from other planets within our solar system (otherwise their atmospheres would be in chemical dis-equilibrium); that soil organisms enhance the rate of rock weathering by at least an order of magnitude (thereby counteracting the long-term, geochemical build-up of carbon dioxide); and the supply of nutrients to life on land depends, in part, on marine algae (e.g. they provide the only known mechanism for replenishing the atmospheric supply of iodine, a trace element essential for animal and human health).

(around 3 million tonnes per year), although now substantially less than anthropogenic sulphur emissions (around 90 million tonnes per year).

Two IGBP components, the International Global Atmospheric Chemistry (IGAC) project and the Joint Global Ocean Flux Study (JGOFS) include research on the dynamics of marine sulphur emissions and their climatic significance. Field work in June 1991 showed that a single, large bloom of the coccolithophorid *Emiliania huxleyi* in the north-east Atlantic (with a total area of around 250 000 km$^2$; Figure 3) released to the atmosphere around 215 tonnes of sulphur per day. That production rate is equivalent to the combined daily emissions of sulphur from seven oil refineries (each processing 20 million tonnes of crude oil per year), or up to ten medium-sized coal-fired power stations

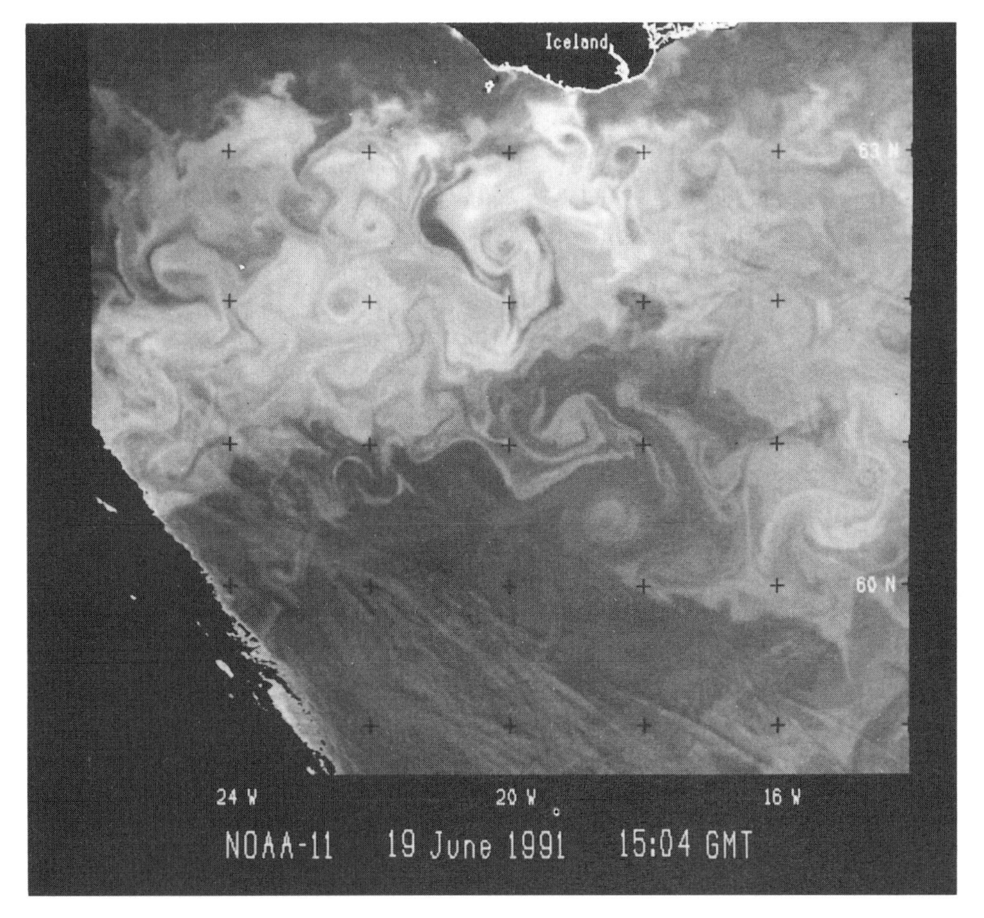


Fig. 3. Satellite image of a bloom of the phytoplankton *Emiliania huxleyi*, south of Iceland. Small, abundant organisms can have large effects: *E. huxleyi* has a global average density of around 1000 individuals per litre in the top 200 m of the ocean, and a single bloom can be responsible for the release of several thousand tonnes of sulphur to the atmosphere, and the precipitation of several million tonnes of calcite to the sea floor. Processing of AVHRR image by S. Groom.

(depending on the sulphur content of the coal, and the efficiency of desulphurization treatments, if any). The pervasive nature of biogeochemical cycles is exemplified by the presence of sulphur 'impurities' in oil and coal, and also by the origins of the calcium carbonate used in flue gas desulphurization. In both cases the sources are biological. Most chalk was deposited by an exceptional abundance of coccolithophorids in the Cretaceous era, around 100 million years ago. In 1991, calcium carbonate production by the *E. huxleyi* bloom was estimated to be 275 000 tonnes per day, exceeding by around three orders of magnitude its production of sulphur compounds.

Are sulphur-releasing algal blooms likely to become more or less abundant if global warming occurs? Applying the Gaia hypothesis, they should increase – with the extra DMS increasing cloudiness, and exerting a cooling, negative feedback effect. However, analyses of polar ice indicate otherwise. In the Vostok ice core from Antarctica (covering the past 160 000 years), the levels of aerosols derived from dimethylsulphide and methane sulphonic acid – products originating from marine algae – were lower during warm, interglacial, periods and higher during ice ages. Whilst there are some problems in the interpretation of these results (since spatial scales for the distribution of the sulphur aerosols are uncertain), they suggest that any cloud-related feedbacks mediated via marine sulphur emissions would enhance climate change, rather than slowing or reversing it.

## Trapped in the Ice

Over time scales of millennia there is strong evidence that an initial, relatively subtle climatic signal can trigger reinforcing changes in the levels of greenhouse gases, providing positive feedback that can then be magnified by other, mostly atmospheric. mechanisms. That interpretation is supported by the analysis of fossil air trapped in polar ice – that shows a remarkably close correlation between temperature changes (assessed isotopically) and the levels of carbon dioxide and methane (Figure 4), with close synchrony between the Arctic and Antarctic. When carbon dioxide and methane were low, glacial conditions prevailed; when they increased, temperatures also rose. Because these gases are relatively well mixed in the atmosphere, the measured levels closely match their global abundance. Confirmation of these relationships is provided by data covering shorter time periods from temperate and tropical mountain glaciers, also by various indirect determinations.

Statistical correlations between changes in atmospheric composition and temperature cannot, on their own, show which is the cause and which the effect, nor do they rule out the possibility that both are responding to an (unknown) 'master' variable. Additional clues are, however, given by the timing of events. Data from the Vostok ice core and elsewhere show that, during warming periods, increases in atmospheric carbon dioxide either lead or are closely in phase with the temperature changes (within the accuracy of

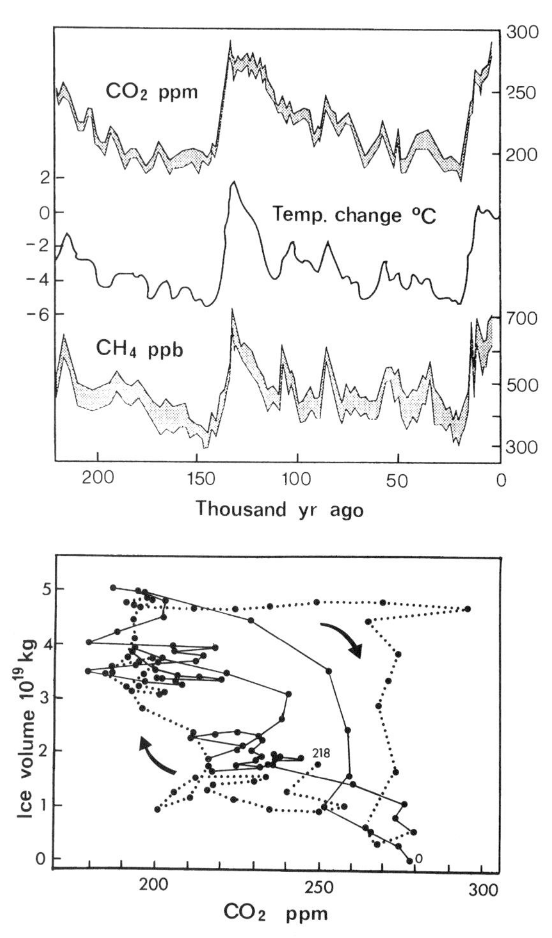


Fig. 4. Relationships between temperature changes and changes in the abundance of the greenhouse gases, carbon dioxide ($CO_2$) and methane ($CH_4$) during two ice-age cycles. Above, data from the Vostok ice core, with temperature estimated from deuterium concentrations and oxygen isotope ratios; below, comparison of the ice core $CO_2$ data with independently derived ice volume estimates, at 2000-year intervals beginning 218 000 years ago. The hysteresis effects shown in this graph are similar for both cycles, and indicate that $CO_2$ increases have greatest potential for positive feedback during their warming phases. After J. Jouzel *et al.*, 1993 (*Nature*, **364**, 407–412) and B. Saltzman and M. Verbitsky, 1994 (*Nature*, **367**, 419).

dating, *c.* 1000 years). However, during the development of glacial conditions, decreases in carbon dioxide concentrations can lag the temperature changes, by up to 4500 years (Figure 4). On that basis, changes in carbon dioxide are more likely to exert a positive feedback effect during warming than during cooling. For methane, the speed of response is independent of the direction of temperature change: both warming and cooling are apparently capable of producing rapid positive feedback. Thus, recent fine-resolution results from the international Greenland Ice-core Project (GRIP) show that, for the period

40 000–8000 years ago, increases and decreases in methane abundance were closely in phase (within 200 years) with the climate fluctuations.

Taken together, the ice age changes in carbon dioxide and methane would have had a direct effect on the Earth's mean radiation budget of about 2 watts per $m^2$, changing mean surface temperature by around 0.7°C. Whilst that accounts for only 15–20% of the isotopically-estimated global temperature change, it would be sufficient to bring additional positive feedbacks into play, acting via water vapour in the upper atmosphere, sea ice, and other changes in radiation reflection and absorption. When these feedbacks are included in current formulations of general circulation models (GCMs), the original change is amplified between 1.5 and 4.5 times. Narrowing that uncertainty range (mainly due to differences in the treatment of clouds within different models) is currently a top priority for deciphering the mechanisms of past environmental change, as well as for future climate projections. Two additional atmospheric forcing factors, not routinely included in GCMs, may also have provided positive feedback during glacial periods: increases in the abundance of cloud condensation nuclei (brought about by variability in marine sulphur emissions, as previously noted), and in dust aerosols, contributing to cooling by reflecting sunlight. Glacial-age ice contains up to 30 times more dust than more recent layers, indicating that cold conditions were also drier and windier. Vegetation maps derived from terrestrial pollen analyses support the view that, at the glacial maximum, there was a considerable expansion of desert areas in continental interiors. However, global-scale atmospheric effects are difficult to quantify: within ice cores, the dust record is spiky, being dominated by particularly strong deposition events.

Analyses of trapped gases, dust content and a wide range of other variables are continuing for the GRIP core, and a second Greenland core (GISP-2). Aspects of particular interest include apparent climatic instabilities during the last interglacial (somewhat warmer than at present) around 125 000 years ago, and the detailed sequence of events during more recent periods of rapid climate change. Both GRIP and GISP-2 are part of the IGBP Past Global Changes (PAGES) project, directed at the collection, synthesis and interpretation of palaeo-data relevant to present and future patterns of Earth System behaviour. In addition to analysing the events of the last ice age cycle, PAGES researchers are constructing a year-by-year record of environmental variability for the past 2000 years, using tree ring analyses, microfossils in varved lake sediments, and other annual layering methods. The aim is to distinguish regional and global variability, and thereby determine the relative importance of short-term solar phenomena (including sun spot cycles), volcanic activity and human actions on contemporary climate dynamics.

## The Carbon Connection

The versatility in the bonding properties of the element carbon makes life possible. It provides the main chemical connection operating within, and transferring energy between, living organisms. There are more than five million known carbon compounds – more than for all the other elements put together. Organic chemists study their structure; biochemists their biological function. Biogeochemists are concerned with a much more limited range of carbon compounds, focusing on those involved in large-scale transfers between the main compartments of the global system: the trade in carbon bonds within the natural economy.

For most of Earth's history, high-energy holdings of carbon have been accrued, through biological processes. The amount of this capital accumulated as fossil fuel is estimated to be at least 5000 gigatonnes (1 Gt = $10^{15}$ g, a thousand million tonnes) of carbon, with a similar amount of organic carbon in soils, surface sediments and dissolved in the oceans. These values greatly exceed the total of around 600 Gt of carbon within living organisms, and the low-energy, inorganic form of carbon ($CO_2$) in the atmosphere, currently around 750 Gt. Total amounts of inorganic carbon are, however, considerably greater, with around 38 000 Gt in the oceans and 90 000 000 Gt in the Earth's crust (much of which has, at some time, passed through living organisms). The idea of a global carbon cycle, connecting these compartments, was first considered by the German chemist Justus von Liebeg in 1840; by the end of the 19th century, several scientists, including Sweden's first Nobel laureate, Svante Arrhenius, had expressed concern regarding the mobilization of carbon by human actions.

The amount of carbon dioxide currently released by burning coal, oil and gas, and (less importantly) from cement production, is known with reasonable accuracy: around 5.5 Gt of carbon per year, in the decade 1980–1990. We also have good data for the rate of its atmospheric increase: about 3.4 Gt per year, for the same period. The amount released by deforestation and other changes in land use is less certain; the Intergovernmental Panel on Climate Change, in its 1994 report, gave a best estimate of 1.6 Gt of carbon per year, with a range of 0.6–2.6 Gt. Other recent estimates have been somewhat higher, at 1.1–3.6 Gt of carbon per year. Whatever the exact amount, it is clear that, at most, only half the anthropogenic carbon entering the atmosphere is staying there. Consequently, there must now be a net uptake, of several gigatonnes a year, in other parts of the carbon cycle – where the *gross* exchanges, due to biological and physicochemical processes, are two orders of magnitude larger, at around 190 Gt per year. Conditions of significant net imbalance in the natural carbon cycle have only previously occurred during the climatic shifts of the ice age cycle. Thus the 'pre-industrial' level of atmospheric carbon dioxide (280 ± 10 ppm) was near-constant from the end of the last glaciation until the advent of large-scale industrial activity and forest clearance, around 1800.

It had been thought that the oceans, with their very large capacity for holding carbon, were mainly responsible for slowing the build-up of fossil fuel carbon dioxide in the air. However, recent studies (based on ocean and atmospheric circulation models, measurements of the bomb isotope $^{14}C$, air-sea exchange studies, and comparisons of the $O_2$ and $CO_2$ changes in the atmosphere) have shown that the current global net uptake of carbon by the oceans is unlikely to exceed 2 Gt (± 0.8) per year. That results in a 'missing carbon' problem, with a quantity equivalent to that produced by deforestation and land use changes unaccounted for. There is no shortage of ideas of where this carbon *might* be going, on land, at sea, or somewhere in between. Suggestions include:

- an increase in the effectiveness of terrestrial carbon sinks, due to regrowth of cleared areas in the tropics;
- a general stimulation of plant productivity on land (particularly in temperate regions of the northern hemisphere), due to the fertilizing effects of increased carbon dioxide;
- additional fertilization effects, both on land and at sea, due to anthropogenic nutrient inputs (e.g. enhanced nitrogen deposition from the atmosphere);
- net accumulation of carbon in the sediments of freshwater systems impacted by human activities, including reservoirs and estuaries;
- a greater-than-estimated uptake in those parts of the oceans (such as polar oceans, coastal zones and shelf seas) where carbon fluxes are complex and poorly represented in global models.

Much research effort is currently directed at these topics, to determine the fate of carbon dioxide derived from fossil fuels, and the global distribution of sinks and sources for other greenhouse gases. There are many associated scientific, geopolitical and ethical problems. For example, will natural processes that are now acting as carbon sinks continue to do so over the next 20–50 years, when further perturbed by human activities that may include significant climate change? Should the national inventories of greenhouse gas emissions (as required by the Framework Convention on Climate Change) include natural and agricultural sources, as well as industrial ones? How much will these emissions have to be cut to achieve 'stabilization of greenhouse gas concentrations in the atmosphere at a level that would prevent dangerous anthropogenic interference with the climate system'? And is it acceptable for developed nations, that have been responsible for most of the rise in greenhouse gases to date, to offset further increases in emissions by enhancing the effectiveness of carbon sinks (e.g. by planting forests) elsewhere in the world?*

All eight IGBP Core Projects include work on the natural dynamics of the carbon cycle, and how such systems may be affected, directly or indirectly, by human impacts. The Joint Global Ocean Flux Study (JGOFS) seeks to quantify carbon transfers within the ocean, and the factors affecting boundary exchanges with the atmosphere, sea-floor and continental margin. Such work is inherently interdisciplinary, involving the study of closely-coupled biological, chemical and physical processes. Primary production by microscopic algae (phytoplankton) in the surface layers is responsible for the 'biological pump', drawing-down carbon dioxide from the atmosphere and contributing to the vertical concentration gradient for dissolved $CO_2$ in the water column (Figure 5).

*See pp 61–62 and 71–75 for discussions of natural and human-enhanced greenhouse gas warming of the atmosphere.

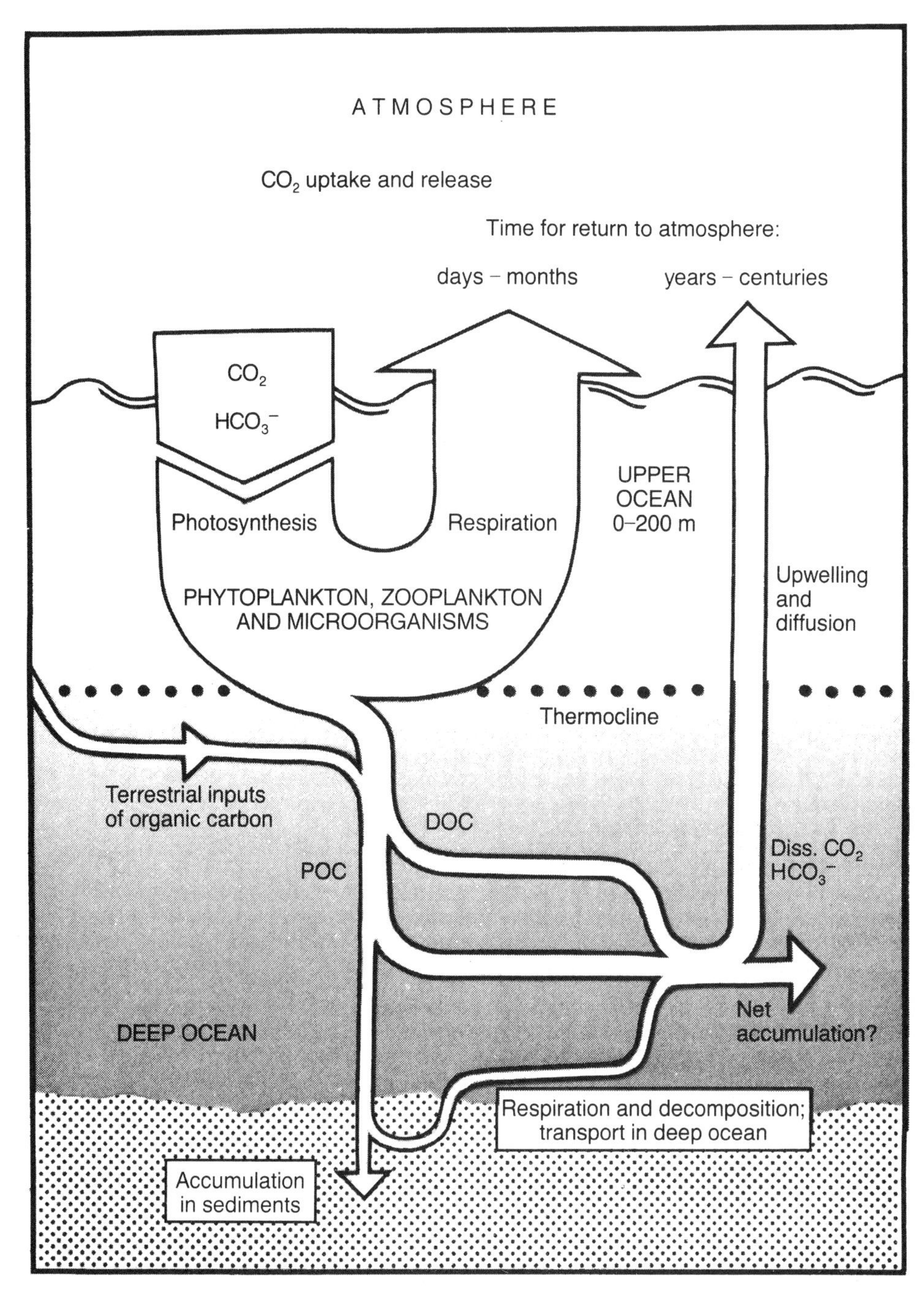


Fig. 5. Biological processes affecting the ocean carbon cycle. Phytoplankton photosynthesis in the sunlit upper ocean lowers the level of dissolved carbon dioxide, causing a drawdown from the atmosphere. Whilst most of the fixed carbon is rapidly returned to the atmosphere, through respiration, some is transferred to deeper water as particulate organic carbon (POC) or dissolved organic carbon (DOC). Return processes may then take years or centuries.

If all marine life ceased, but ocean circulation was unaffected, it is estimated that there would be an increase of at least 50% in atmospheric carbon dioxide, from current levels of 350 ppm to around 540 ppm.* However, if, in addition, present-day circulation patterns were also disrupted, there could be further increase of similar magnitude, raising the atmospheric total to around 700 ppm. The latter effect would result from the shut-down of the physico-chemical 'solubility pump', with deep-water transfers from high to low latitudes. The controls of these processes are being investigated by JGOFS researchers, with special emphasis on the North Atlantic, Equatorial Pacific, Arabian Sea and Southern Ocean. Wider space and time perspectives are provided by global surveys, satellite data, time series studies and analyses of palaeo-records (with links to PAGES). When the controls of the ocean carbon cycle are better known, and improved ocean circulation models are available (through the WCRP World Ocean Circulation Experiment, WOCE), it should be possible to provide reliable simulations of oceanic responses to future climate change – taking account of altered ocean circulation patterns and their effects on nutrient distributions and marine productivity.

The dynamics of land–sea carbon transport (mostly via rivers) and the vertical and lateral carbon exchanges in coastal regions are major themes for the Land–Ocean Interactions in the Coastal Zone (LOICZ). There is joint work with JGOFS at the edge of the continental shelf, and with the Biospheric Aspects of the Hydrological Cycle (BAHC)

**Box 4. Algae, iron and the ocean carbon sink**

IGBP research is *not* directed towards the development of technological fixes, to counteract the consequences of man-made global change. Nevertheless, the programme does have a role in assessing the scientific validity of potential mitigation strategies, and for examining their wider environmental implications. One suggested method to slow the build-up of atmospheric carbon dioxide is to enhance its ocean uptake, by stimulating primary production in surface waters. This idea was developed from bottle experiments, showing that for the Southern Ocean and much of the Pacific (where major nutrients are abundant, but productivity is low), small additions of iron can dramatically increase phytoplankton growth.

Field experiments to test the iron hypothesis were carried out in the Equatorial Pacific in October 1993 and June 1995. The international research team found that around half a ton of iron, spread over 50–100 $km^2$, was sufficient to initiate an algal bloom, with a rapid order of magnitude rise in chlorophyll and plant biomass, and associated increases in aerosol-forming sulphur emissions. Effects on carbon dioxide uptake varied between the two experiments, and much of the new plant growth was apparently subject to rapid recycling. On the basis of these results, iron fertilization of the open ocean provides a very useful scientific tool, for 'turning-on' primary production, but it should not be relied upon for the control of atmospheric carbon dioxide.

*In contrast, atmospheric oxygen concentrations would hardly be affected, because of the natural abundance of oxygen in the atmosphere.

project regarding carbon fluxes in river basins. Atmospheric budgets for carbon dioxide are addressed by the International Global Atmospheric Chemistry (IGAC) project, principally through a global network of lower-atmosphere $CO_2$ monitoring stations (jointly supported by the World Meteorological Organization). Knowledge of atmospheric distribution patterns, together with data from associated intercalibration and isotope studies, provides the basic input for global carbon cycle models; such information also helps to identify (and locate) source and sink processes.

On *terra firma* the newly established Land Use/Cover Change project (LUCC) will analyse and quantify the socio-economic driving forces for deforestation, afforestation and other land management changes that affect carbon fluxes. Many of the LUCC studies will be closely linked to those of the Global Change and Terrestrial Ecosystems (GCTE) project; for example, the GCTE experiments on the effects of enhanced $CO_2$ (under open air conditions, Figure 6) for both managed and natural ecosystems, and the development of global, dynamic vegetation models. These models now include a range of $CO_2$-related feedbacks, and can be coupled, in a preliminary way, to the physical climate models of the World Climate Research Programme. There also promises to be a productive three-way interface between GCTE, LUCC and the Past Global Changes (PAGES) project, using reconstructed land cover patterns to examine the relationships between vegetation and different climate regimens.

Early results of the GCTE studies show the complexity of the responses of terrestrial ecosystems to changes in atmospheric composition. As atmospheric carbon dioxide rises, fertilization effects will undoubtedly occur. However, there are many other interactions with temperature and seasonality, soil nutrients and water, plant nutritional quality, biodiversity, and pests, weeds and diseases. Whilst equilibrium climate/ecosystem models

Fig. 6. Will extra carbon dioxide in the atmosphere enhance the productivity of crops and natural ecosystems? These questions are being investigated by IGBP using long-term, free-air circulation experiments, as shown here for a cotton field in Arizona. Photo: J. R. Mauney.

for doubled $CO_2$ conditions show a significant increase in terrestrial carbon storage, those simulating the transient response indicate that additional releases of carbon dioxide (from a greater incidence of wildfires, and enhanced respiration in tropical soils) may exceed storage responses during the next 50–100 years. Claims have been made by some agriculturalists that increases in atmospheric carbon dioxide will be overwhelmingly beneficial, increasing crop yields by 30% and bringing about a 're-birth of the biosphere'. Manipulative experiments on natural ecosystems support a more cautious view, and show the importance of long-term studies. For example, tussock tundra in the Arctic initially showed a strong growth response to doubled $CO_2$, but that effect declined over successive seasons. After 3 years there was no effect at ambient temperatures, although net fertilization did still occur when (in addition to doubling $CO_2$) temperatures were raised by 4°C.

## Hydrological Drivers

Water would be a more appropriate name for our planet than Earth. Its liquid and solid forms together cover three-quarters of the world's surface; as a gas and aerosol it has more effect on the global climate than any other atmospheric constituent; and its restless movements, on land and at sea, shape the landscape (ultimately moving mountains) and carry heat and chemicals around the world. Water is also the medium within which life first evolved, and its molecular and ionic forms are as essential as the element carbon for the continuity of life. As recognized by Leonardo da Vinci, 'Water is the driver of nature', with very special importance to the Earth System.

Living organisms have developed different physiologies according to whether they inhabit (or drink) an aqueous medium containing appreciable amounts of dissolved salts, or one that has been recently distilled, by evaporation and precipitation processes. Most terrestrial species, including ourselves, depend on the latter – freshwater – that, as a liquid, can be in short supply, despite water's overall abundance (see Box 5). Thus rainfall generally has a greater effect than temperature in determining the distribution and abundance of life on land. Furthermore, the regional variability in freshwater delivery, and the widespread occurrence of temporal events (droughts and floods), make it much harder to analyse the causes and effects of changes in global precipitation patterns than for temperature.

Research priorities relating to the future human use of water resources were reviewed at ASCEND 21 (Dooge *et al.*, 1992), an ICSU meeting held in preparation for the 1992 UN Conference on Environment and Development. There were two main concerns: first, that freshwater use in many arid and semi-arid countries already exceeds its natural rate of re-supply. Aquifers are being drained that had taken thousands of years to fill, with their future capacity for replenishment reduced by consolidation of the water-bearing strata. Second, our understanding of hydrological processes is largely based on work in the temperate zone, where evaporation rates are generally low and capital-intensive

**Box 5. Water, water everywhere?**

The size of the components of the global water distribution can be expressed in terms of average depths if equally distributed over the entire Earth. Saltwater oceans and seas would have an equivalent depth of around 2750 m (97% of all water); polar ice caps and glaciers, 60 m (2.1%); groundwater, 20 m (0.7%); rivers and lakes, 26 cm (0.01%); soil moisture, 4.8 cm (0.002%); and atmospheric water, a paltry 2.8 cm (0.001%). The average daily rain and snowfall, expressed similarly, is 0.3 cm. Clearly there is no risk of water shortage for marine organisms, but for those on land, freshwater may be in short supply. Being close to the sea may provide a wet, maritime climate, but not necessarily. One of the world's rainiest places is Mt Waialeale, on Hawaii, where the annual average downpour is 1170 cm; however, one of the driest is also within sight of the Pacific, at Arica, Chile, where there was a 14-year absolute drought earlier this century, and the long-term average rainfall is only 0.8 mm per year.

Human needs for freshwater are at present poorly met. The World Bank estimates that 80 countries now have shortages that threaten health and economies, and 40% of the world's population (more than two thousand million people) have no access to clean water or sanitation. Since 1900, the human use of freshwater has increased from around 500 $km^3$ per year to around 2500 $km^3$ per year, and, at the current rate of increase, usage will double within the next 20 years. However, supplies cannot keep up with such demand, and increasing scarcities are inevitable – regardless of whether climate change might alter rainfall patterns.

engineering has, to date, ensured that water is usually plentiful (sufficiently so that, in the USA, up to 4500 litres of water may be used to produce a 200 g steak). In the tropics and in the dry subtropics, the interactions between water, soil and vegetation are very different, and are being changed in ways that are not sustainable; see Figure 7, for example. Thus, many large dams are filling with sediment much more rapidly than expected, and the benefits of water management schemes in semi-arid areas are frequently proving to be short-lived. For example, in the Sahel there have been irreversible reductions in soil fertility for many irrigated croplands, due to their increased alkalinity and salt content; in West Bengal around 200 000 people now suffer from cumulative arsenic poisoning, as a result of their recent dependence on pumped, mineral-rich groundwater (with contamination levels rising as the water table falls).

Hydrological processes are of central importance to general circulation models (GCMs), being closely linked to energy exchanges in the atmosphere, at the land surface, and at the air–sea boundary. Unfortunately GCMs cannot yet provide realistic simulations of the global water cycle: in addition to the high short-term variability of natural weather patterns, the horizontal resolution within the models (typically 300–1000 km) is too large for inclusion of critical topographical features affecting water fluxes. Thus, local precipitation extremes (e.g. at Hawaii and Arica, see Box 5) are smoothed over, and there is a poor match between model results and measurements for present-day flows of major

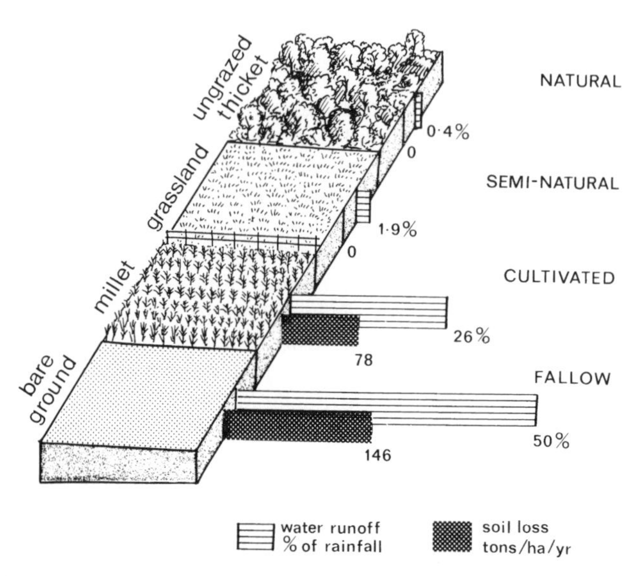


Fig. 7. Effect of vegetation cover on rainwater runoff and soil erosion: results of a 2-year study in Tanzania. Data from A. Sundborg and A. Rapp, 1986 (*Ambio*, **15**, 215–225).

rivers. Furthermore, when different models are compared under climate change scenarios (such as doubled $CO_2$), there are large regional differences regarding their predicted changes for hydrological parameters, with no agreement on how global warming would affect soil moisture in tropical and semi-tropical regions – a key issue for agricultural productivity, ecosystem structure and feedback responses. More powerful computers (using a smaller grid size), with better treatment of cloud dynamics, should help to resolve such issues. However, it is also necessary to improve the simulation of many water exchange processes occurring at the land surface, where biology, physics and chemistry interact.

Although most atmospheric water vapour originates from (and directly returns to) the sea, there is also extensive re-circulation over land. This precipitation source is particularly important for continental interiors, such as central Asia, the American mid-west, and the Amazon basin. Soil properties have a major effect on water absorption and retention, whilst plants extract groundwater and lose it through their leaves (by evapotranspiration, a cooling process), thereby making it more likely that rain will fall again on land rather than at sea. Very large amounts of water pass through vegetation each year, around 30 000 $km^3$, equivalent in magnitude to the total flow of all the world's rivers. Where either soil or plant cover is reduced or lost from a small area (e.g. by local erosion or deforestation), greater, but more erratic, river run-off usually occurs. This has long been recognized: 'Destructive torrents are generally formed when hills are stripped of the trees which formerly confined and absorbed the rains' (Pliny, quoted by George Perkins

Marsh in *Man and Nature*, 1864). However, larger-scale effects are more uncertain, and it is only within the past decade that soil–vegetation–atmosphere transfers of water have been studied in a quantitative and comprehensive way, using remote sensing and other global databases to assess their climatic significance. Models describing the water/climate relationship must take account of a wide range of environmental variables, including solar energy absorption effects (albedo); the influence of plant structure on near-ground wind strength; the intimate links between water and carbon dioxide fluxes; and the many effects of living – and dead – organisms on water run-off and other soil properties (Figure 8).

The IGBP project Biospheric Aspects of the Hydrological Cycle (BAHC) provides international coordination for such studies. Models of the water cycle are being developed that relate freshwater availability to climatic and biogeochemical processes (Figure 9), working closely with other IGBP projects (particularly GCTE and IGAC) and the WCRP Global Energy and Water Cycle Experiment (GEWEX). In addition to scaling-up the physiological and ecological processes that link water, energy and carbon fluxes, BAHC researchers are also 'downscaling' GCMs, to improve the local match between

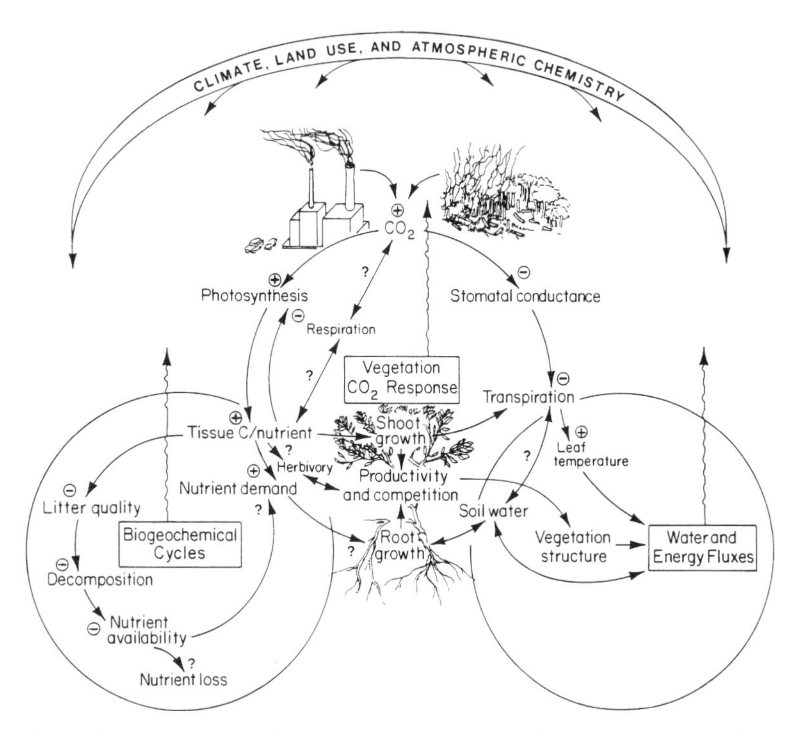


Fig. 8. Effects of changes in carbon dioxide levels on vegetation, and hence on nutrient cycling, and water and energy exchanges. Plus and minus signs indicate potential positive and negative feedbacks, respectively. From the *GCTE Operational Plan*, IGBP Report 21, 1992.

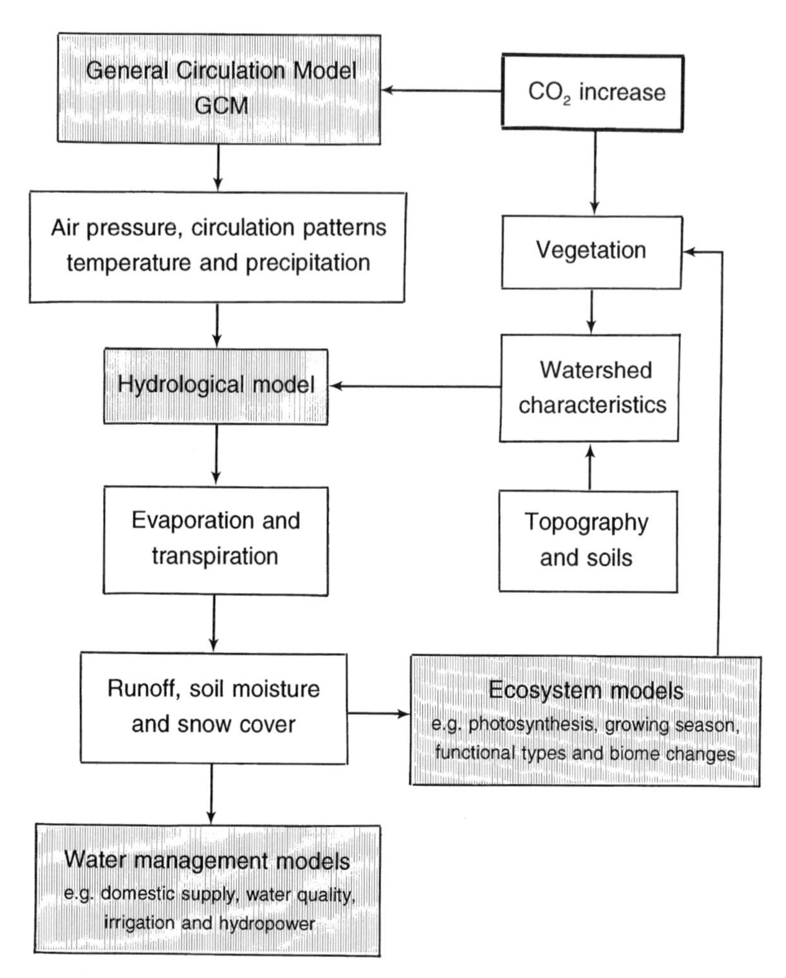


Fig. 9. The linkages between climate models (GCMs), hydrological models and ecosystem models that are needed for realistic regional-scale assessments of freshwater availability, and the sensitivity of such resources to global change impacts. Based on the *BAHC Operational Plan*, IGBP Report 27, 1993.

model data and the real-world features of topographical and temporal variability. The role of rivers in the lateral transport of carbon, nutrients and other biogeochemically-important materials connects BAHC and the Land-Ocean Interactions in the Coastal Zone (LOICZ) project, recognizing that natural events and human impacts far from the sea can have major effects on estuaries, deltas and coastal ecosystems. There are large differences between the properties of the relatively well-studied (and usually slow-flowing) river systems of the temperate zone, and the behaviour of more dynamic systems, as found in geologically-young mountain regions, semi-desert regimens, and island systems of the humid tropics (Figure 10). Sediment transport by ephemeral desert rivers can be as much as 400

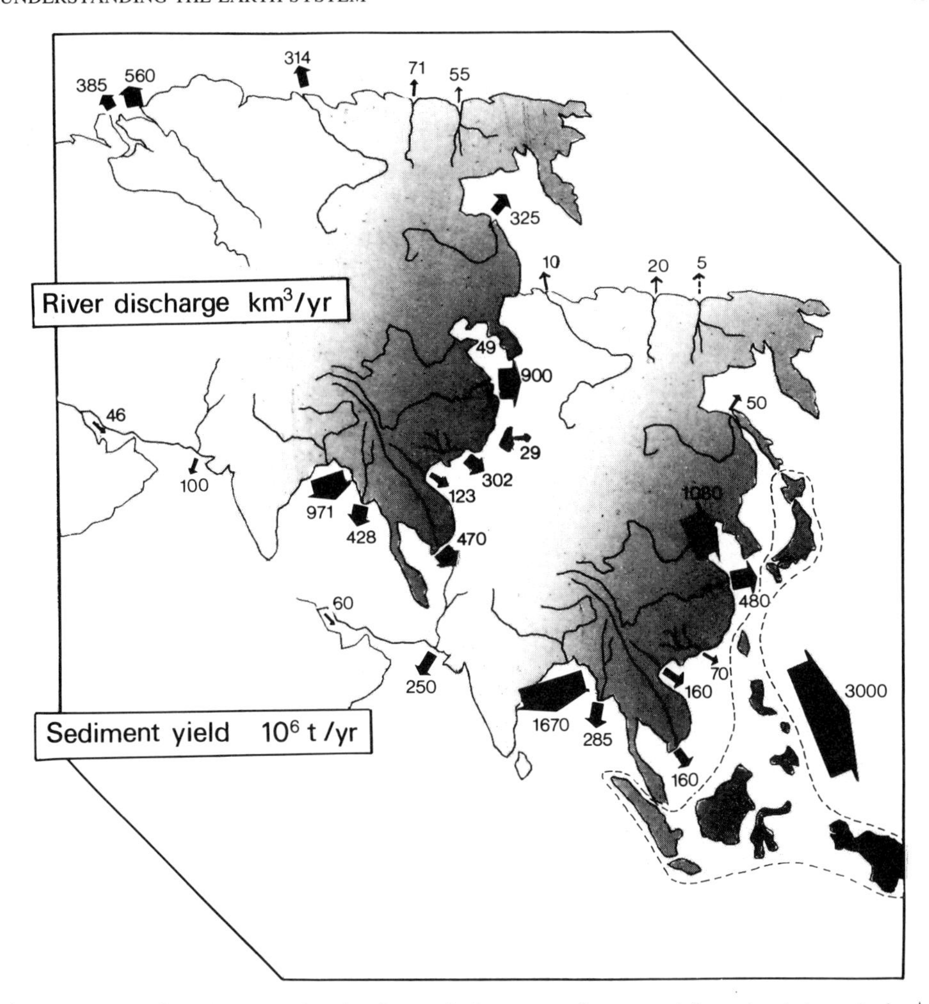


Fig. 10. Water flows (upper map) and sediment discharge rates (lower map) for major drainage basins in Asia. Sediment transport/water flow ratios are very much higher for tropical and semi-tropical rivers than for those draining into the Arctic – with important implications for water management (e.g. hydropower schemes), coastal erosion and marine productivity. Data from J. D. Milliman, 1991 (in *Ocean Margin Processes in Global Change*, ed. R. F. C. Mantoura, J.-M. Martin and R. Wollast).

times higher per unit volume of water (although for only a few days a year) than that of their perennial counterparts.

The LOICZ project also addresses the direct influence of ecosystem processes and human impacts on coastline structure, including the consequences of the increasing exploitation of coastal resources – such as the draining of saltmarshes, felling of mangrove forests and destruction of coral reefs. In highly populated river delta systems the effects of basinwide and locally operating human activities are combined, frequently causing a dramatic fall in land levels (e.g. as a result of a lack of sediment re-supply, natural consolidation, and groundwater extraction). Such processes can cause a relative sea level rise of 1–2 cm per year, compared with the current global rise in absolute sea level of 1–2 mm per year. The latter change in sea level is usually ascribed to the thermal expansion of ocean water, due to this century's warming trend; however, worldwide human impacts may also be involved. Whilst reservoirs increase water storage on land, groundwater withdrawal and land-use changes promote its transfer from land to sea. The net effect of these anthropogenic influences is uncertain: recent estimates range from an increase of 0.5 mm per year in global sea level, to a decrease of 1.6 mm per year. Resolution of this issue is clearly of great importance for projections of sea level rise over the next 50–100 years.

## Up in the Air

When carbon dioxide is in the atmosphere, its arithmetic is relatively straightforward. Thus, it is added, subtracted and carried over; the few chemical transformations that also occur are of little importance. However, other climatically-important atmospheric constituents are much more reactive, and their changes this century have affected the Earth's heat budget at least as much as changes in carbon dioxide. To predict future changes in radiative forcing, it is therefore necessary not only to understand the processes affecting the strength of sinks and sources of the full range of greenhouse gases, but also to work out their interactions in the chemical cocktail above our heads (Figures 11 and 12).

Methane ($CH_4$) is of particular interest. This gas is around 20 times more powerful, molecule-for-molecule, than carbon dioxide in its greenhouse properties, and its involvement in past climate changes has already been mentioned. During the 1970s and early 1980s, atmospheric methane increased rapidly, at around 1% per year; however, recent increases have not been so great. All significant sources of methane (12 are identified by IPCC) are directly or indirectly of biological origin. Whilst around 100 million tons of 'old' methane are released annually by human activities (from the coal, natural gas and oil industries), around four times that amount is freshly produced by biomass burning and by bacteria in oxygen-poor environments. The latter include natural wetlands, rice paddies, and inside the guts of cattle and termites. Some methane from the air is broken down by methane-consuming microbes in the soil and upper ocean, but around 95% of the losses occur through chemical reactions in the atmosphere – principally through its oxidation by the hydroxyl radical (OH).

The hydroxyl radical also reacts with other trace components of the atmosphere, including many of those added to the air by industrial activities; for example, more complex hydrocarbons, sulphur dioxide and nitrogen oxides. As a result, it is very short-lived, with an average lifetime of around one second between its photochemical formation (involving ultraviolet light) and its breakdown through scavenging oxidations. Levels of OH cannot be easily measured; however, calculations suggest that its availability declined by about 25% in the period 1950–1985. Current and future trends are uncertain.

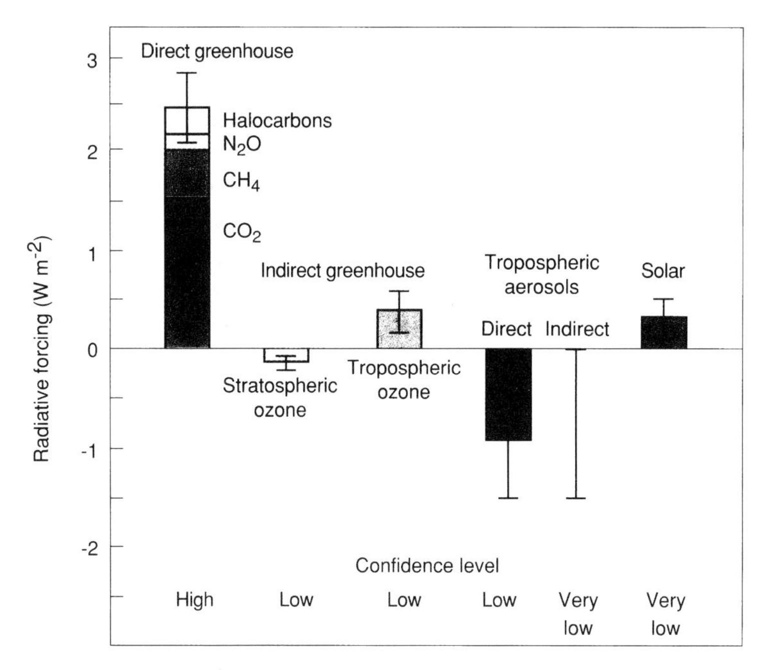


Fig. 11. Estimates of globally averaged warming and cooling effects due to changes in atmospheric composition and solar activity since *c.* 1850. From IPCC, 1994; *Radiative Forcing of Climate Change.*

Much of the change in atmospheric levels of methane, as noted above, could therefore have been due to variability in the availability of OH – with that, in turn, affected by general changes in pollutant levels.

Progress in resolving such complex problems demands sophisticated experimental work, together with a global exercise in gathering, integrating, and interpreting data covering a wide range of research topics. The International Global Atmospheric Chemistry (IGAC) project provides the necessary framework for such action, through a combination of field observations, laboratory studies and theoretical work. IGAC effort is directed at five regional foci (marine, tropical, polar, boreal and mid-latitude), a global focus and a fundamental focus. An additional focus on aerosols was added in 1995, following the merger of the project with the International Global Aerosol Programme (IGAP). Close links with other IGBP projects are essential, since atmospheric composition is intimately connected with events at the land and sea surface. Thus, as well as investigating decadal trends and micro-second reactions in the atmosphere, IGAC addresses such diverse phenomena as biomass burning in African savannas; microbial metabolic rates in Arctic peat bogs; the effect of land use changes on trace gas fluxes in southeast Asia; and the processes affecting air–sea gas exchange rates and aerosol formation, with links to upper ocean productivity and the atmospheric deposition of nutrients.

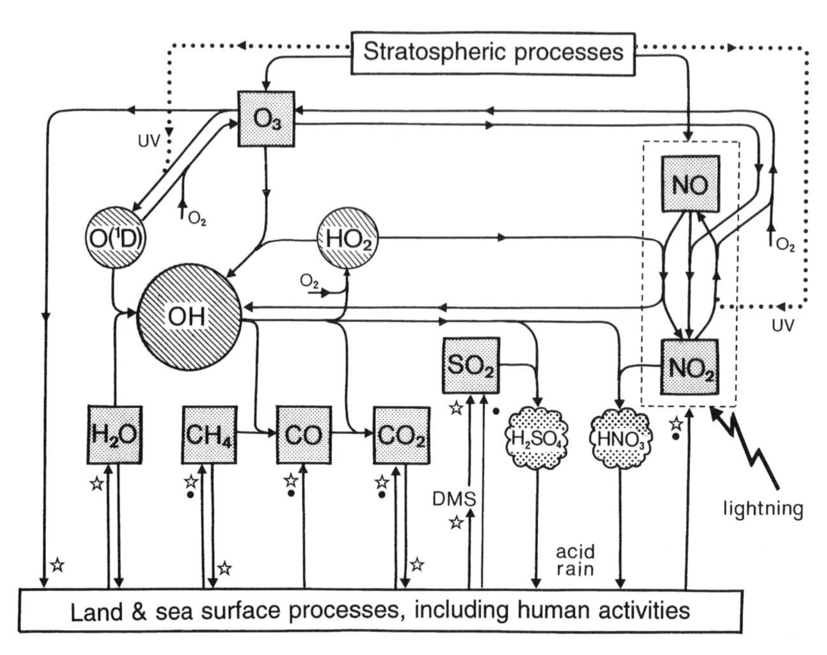


Fig. 12. An outline of key chemical reactions in the lower atmosphere relevant to the Earth's heat budget. In boxes, greenhouse gases and pollutants; in circles, reactive radicals; in clouds, reflective aerosols. Sources and sinks: ☆, biospheric (including agricultural); ●, industrial. Most greenhouse gases also have volcanic sources, and other exchange processes associated with the hydrological cycle and ocean solubility changes. Based on R. G. Prinn, 1994 (*Ambio*, **23**, 50–61).

The atmosphere also provides the link with external energy sources, and IGAC research contributes to, and benefits from, the development of physically-based climate models within the World Climate Research Programme, e.g., the recent inclusion within GCMs of the effects of man-made and natural sulphate aerosols, that promote cloud formation and thereby counteract global warming by greenhouse gases. These effects are strongest in the Northern Hemisphere, where regional reductions in solar radiation are estimated to be as high as 2 watts per $m^2$ (Figure 13). However, there is no single value for the cooling caused by a specified quantity of sulphur emissions. That is because the formation of sulphuric acid aerosols from sulphur dioxide involves the OH radical – linking it to virtually all other atmospheric reactions – and the aerosols themselves are relatively short-lived, lasting just a few days. Furthermore, the reflectance properties of clouds are affected by other, combustion-related, atmospheric constituents. Thus, a cloud containing water droplets darkened by soot may cause warming, whilst a clean, white one is much more likely to result in cooling. The spatial and temporal distribution of different emissions is therefore of critical importance: the faster the oxidation rate of sulphur dioxide, the smaller the region over which sulphuric acid particles are produced – and hence, in general, the smaller the effects on incoming solar radiation. See also p. 87 and Kuemmel (1996).

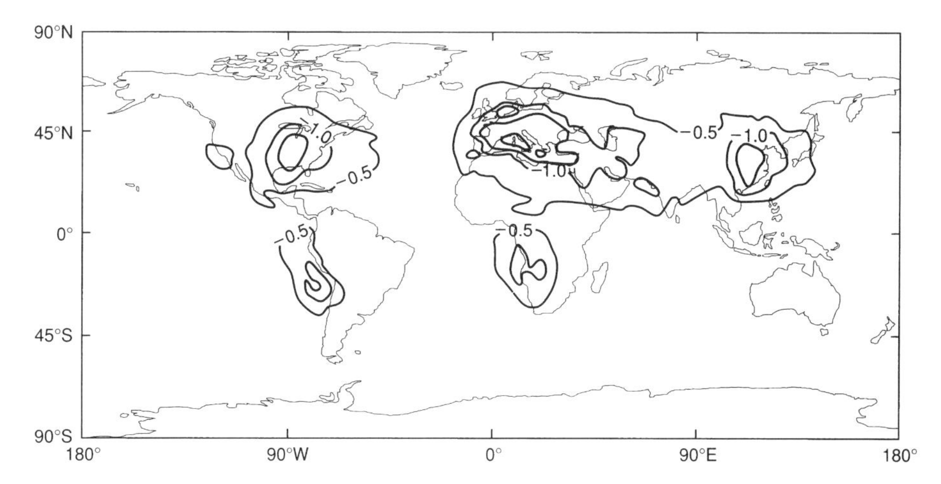


Fig. 13. Cooling effects of current loadings of sulphate aerosols, added to the atmosphere through human activities. Radiative forcing (watts per $m^2$) is greatest in the northern hemisphere, over or close to industrial regions. From IPCC, 1994; *Radiative Forcing of Climate Change*.

**Box 6. 1992: an eventful year**

It will be some time before the success of the 1992 United Nations Conference on Environment and Development (UNCED) can be fully assessed. It may also be some time before we can satisfactorily interpret several unusual changes in atmospheric chemistry occurring in that year. At least four greenhouse gases were affected: carbon dioxide, methane, nitrous oxide and carbon monoxide.

For carbon dioxide, there was an exceptionally low increase (0.2%; 1.5 Gt C) in its atmospheric abundance in 1992, less than half its rate of rise for 1991 and the average for the previous decade (of around 0.5% per year). For methane, the annual rate of increase was already in decline, from about 1.3% in the late 1970s to about 0.6% per year in 1989. In late 1991 the rate of increase slowed further, and, for the Northern Hemisphere, there was no increase at all in 1992. Some reductions in anthropogenic emissions of methane may have been made (due to economic factors affecting fossil fuel use and deforestation rates, or to reduced methane leakage from natural gas pipelines); however, such savings are estimated to be at least an order of magnitude too low to account for the observed atmospheric effects.

Could the eruption of Mt Pinatubo in June 1991 have been a contributing factor? Mechanisms have been suggested whereby this event indirectly increased the oxidizing capacity of the lower atmosphere (e.g. by increasing the availability of OH, linked to reduced stratospheric ozone and enhanced UV penetration). That might explain not only the methane changes, but also the reduced rate of increase in nitrous oxide, and the significant fall (by 18% between June 1991 and June 1993) in carbon monoxide levels. The decreased rate of carbon dioxide accumulation cannot be due to that effect; however, it has been speculated that the Mt Pinatubo eruption may have caused a net global stimulation of photosynthesis, due either to cooling effects (and increased rainfall) or to an increased supply, via volcanic dust aerosols, of micro-nutrients (such as iron) to the open ocean.

At present, the global warming (or cooling) potentials of atmospheric constituents are assessed according to their individual characteristics. Such descriptions are unsatisfactory as predictive tools, since they do not take account of interactions affecting their atmospheric lifetimes and radiative properties, as indicated above. The alternative approach, being developed by IGAC, is to describe the behaviour of the ensemble through computer packages. Using that method for assessment exercises (e.g. by IPCC), and for formulating control strategies (e.g. through the Climate Convention) will present problems that are complicated but not intractable. However, it will mean a departure from the previous practice of considering all greenhouse gases in terms of their carbon dioxide equivalents, and increases the complexity of making cost/benefit analyses within the overall goal of sustainable development. The environmental benefits of reducing acid rain and protecting ozone levels in the upper atmosphere were previously regarded as self-evident – but it now seems that such measures could enhance global warming to some degree, by reducing aerosol reflectance and slowing the rate of OH production. A sound internationally accepted knowledge base is essential for all areas of global change research – but particularly so at the 'front line' of atmospheric chemistry.

## Bringing It All Together

In companies and government, major financial decisions are nowadays likely to be made on the basis of simulation models. They are also increasingly used as the principal tool for handling information-related problems, to assist in developing strategies, setting targets and guiding the systematic treatment of risks. Economic models are fallible, and can be highly sensitive to initial assumptions and unforeseen events; nevertheless, they provide a conceptual framework, procedures for the manipulation of large and complex data sets, and insights into system behaviour that cannot be obtained by other means. Earth System models share those generic attributes, and are being developed (without the restrictions of commercial confidentiality) for closely similar purposes in relation to the natural economy. One additional role must be stressed: scientific understanding. If relationships can be quantified as a function of fundamental properties – 'natural laws', rather than just as an empirical, best-fit solution – there can be much more confidence when those relationships are used to explore scenarios beyond the bounds of direct observation.

No model, however sophisticated, can exactly mimic the real world. The goal is not to be as complicated as possible, but to include all aspects that are essential for a reasonably realistic simulation of whole-system behaviour. Such internal interactions include the possibility that gradual changes in one parameter may produce abrupt jumps (non-linear or chaotic behaviour) in other features. Many uncertainties in whole-system responses to external forcing can be reduced by better descriptions of component reactions. Whilst other, new responses may also arise, well-constructed models can provide information on the probability of such events. The overall need is to replace conditions of uncertainty (when a wide variety of optimistic or pessimistic scenarios seem equally likely), with knowledge of risks, and their statistical properties.

Global model-builders recognize that there is considerable scope for improving the reliability of the products of GCMs, both for best estimates and the range of possible values. It is conceivable that the inclusion of biogeochemical feedback effects will make little overall difference to GCM model behaviour. However, effects could also be considerable – as indicated by preliminary simulations of vegetation–climate interactions (Figure 14). Too much is at stake not to make the effort. The global modelling of the physical climate system was developed from computer-based weather forecasting; it has a very large observational data set, and its methodologies have evolved over more than thirty years, increasing in sophistication. Modelling of the global biogeochemical system is a more recent endeavour, and is only now completing its first phase – the identification of component biospheric interactions that have most effect on the system as a whole.

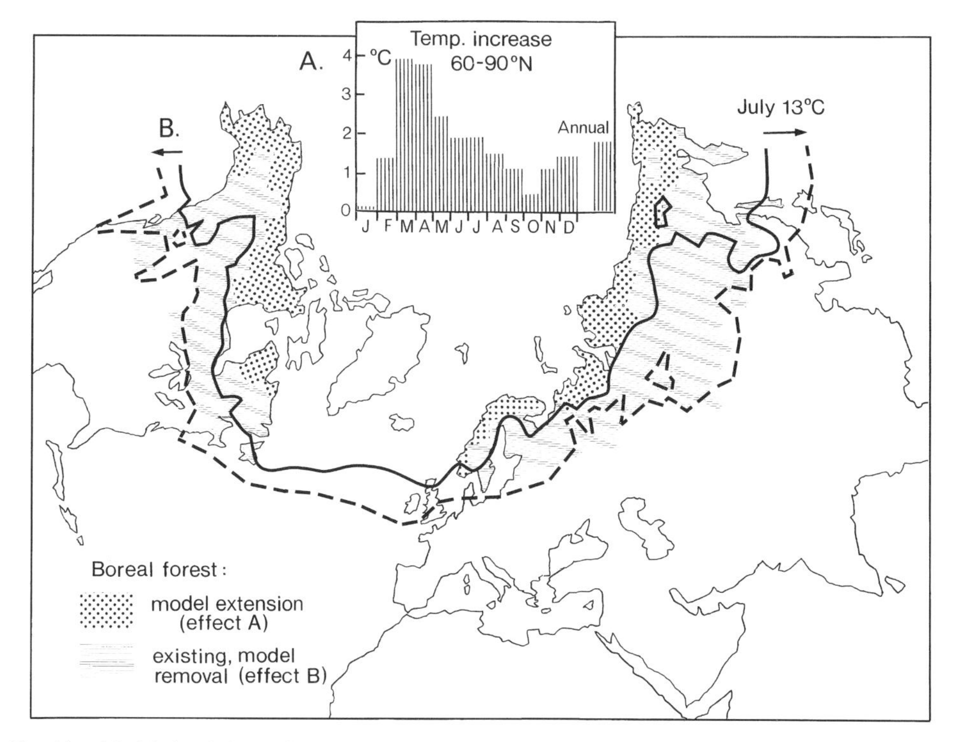


Fig. 14. Model simulations of the dynamic interactions between high latitude vegetation and climate. **A**: Warming effect of imposed forest expansion (to match the estimated distributions of *c*. 6000 year ago) on zonal-average land surface temperatures, between 60 and 90°N. **B**: Cooling effect of the simulated removal of current boreal forest cover, causing a southward displacement of 1000–2000 km for the 13°C isotherm for July. After J. A. Foley *et al.*, 1994 (*Nature*, **371**, 52–54) and G. B. Bonan, D. Pollard and S. L. Thompson, 1992 (*Nature*, **359**, 716–718).

Relatively simple models can be used initially, provided their limitations are recognized and major sensitivities identified. Such models are easier to design and understand, and can be readily integrated over long periods.

Having shortlisted the potentially important sub-systems, it is necessary to improve the formulations of their behaviour, validating model descriptions with experiments and observations over as wide a range of conditions as possible (e.g. using palaeo-data), thereby progressively increasing the complexity of different couplings and combinations. This hierarchical, step-by-step approach is a key feature of modelling within IGBP, with the aim of full partnership with the modelling efforts of the World Climate Research Programme. Each IGBP Core Project is therefore developing its own suite of basic models. In addition, IGBP has established a cross-cutting activity on Global Analysis, Interpretation and Modelling (GAIM), to ensure model complementarity (for linking to other sub-systems) and to assist in identifying knowledge gaps. Initial attention is being given to current integrating work on the carbon cycle: its present day features, its changes during the past 200 years (the 'fossil fuel era'; Figure 15), and its status at key times during the past 150 000 years. Since both carbon dioxide and methane are being considered, these modelling studies require information on the dynamics of a large range of

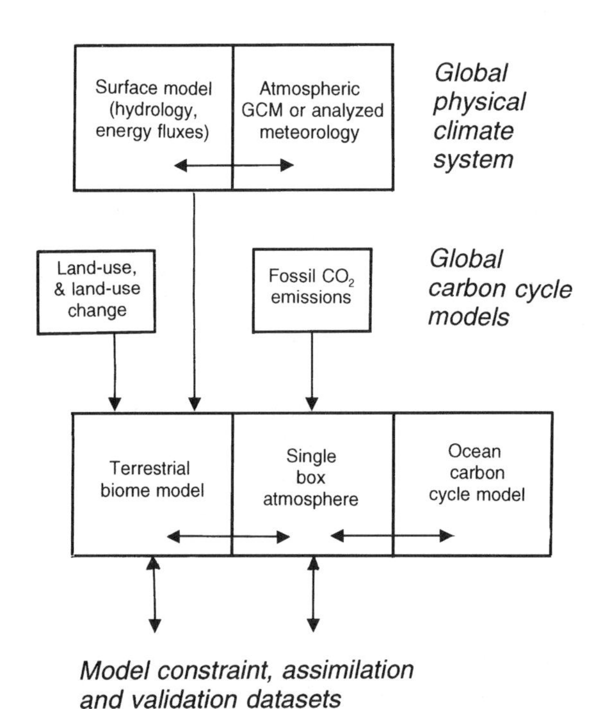


Fig. 15. Inter-model linkages that are being developed through the GAIM study of interactive changes in terrestrial carbon storage, 1700–1990. Based on *IGBP Global Modelling and Data Activities 1994–1998*, IGBP Report 30, 1994.

terrestrial, marine, atmospheric and hydrological processes – with the necessary data and process descriptions being provided by the IGBP Core Projects.

Global biogeochemical models also require the observational equivalent of world weather reports: basic data sets that describe, with reasonable geographic precision, the ongoing status and seasonality of terrestrial and marine ecosystems, and the distributional patterns of key features of atmospheric chemistry. As a consequence of space exploration (and arguably one of the most important benefits to humanity from that activity), our ability to map such properties of the Earth's surface has increased remarkably in recent years. Many further improvements are expected by the turn of the century. The amounts of data provided by satellite remote sensing are awesome. In 10 years the Coastal Zone Colour Scanner provided 1 terabyte ($10^{12}$ units of information) of data, which had a dramatic effect on our understanding of ocean processes. Plans for the suite of satellites comprising NASA's Earth Observation System (EOS) envisage data deliveries of 2 terabytes per day, for 15–20 years from the late 1990s.

The IGBP Data and Information System (IGBP-DIS) has been established to maximize the usefulness of that information supply to IGBP, both for GAIM modelling and for individual projects. IGBP-DIS has played a major part in developing an on-going, weekly compilation of data on global land-surface characteristics, at 1 km resolution, derived from weather satellite sensors (AVHRR: Advanced Very High Resolution Radiometer). This provides information on many important attributes of vegetation cover, including leaf area index, leaf biomass, net primary productivity and photosynthetic capacity. Other global databases being developed by IGBP-DIS, from both satellite and non-satellite sources, include those on biomass burning (linked, via IGAC, to data on fire emissions); the distribution and key properties of wetlands; and soil characteristics, extending existing classifications to include such features as carbon content and water-holding capacity. For all those activities, IGBP-DIS works in close coordination not only with IGBP Core Projects but also with space agencies, international data centres and the many other bodies involved in the gathering, handling and management of environmental data.

As indicated above, observations from space are invaluable for Earth System research. But interpretation of those data streams (and knowing their strengths and limitations) requires a well designed suite of *in-situ* studies, that are also necessary for the detailed investigation of biogeochemical processes. The many experiments and measurements needed for IGBP research must be carried out on a worldwide basis by individual scientists, not committees. Nevertheless, there is added value, more rapid progress and a greater chance of overall success when effort is not only coordinated nationally, but also at regional and international levels. The framework provided by national membership of IGBP (Figure 16) greatly facilitates the global integration of data and knowledge; it can also assist in the formulation and application of sound policy responses to global change problems. For those reasons, IGBP has given high priority to regional collaborations and networks – to encourage the full involvement of both developed and developing countries

Fig. 16. Geographical distribution of the 73 National IGBP Committees, August 1995.

in global change research, and the sharing of scientific and technological expertise. These initiatives are being developed jointly with WCRP and HDP, through the System for Analysis, Research and Training (START).

With scientists in over seventy countries working together, major advances in understanding our changing planet can be expected within the next ten years. It is already apparent that many highly important features of the ever-changing Earth System arise from the interactions of its components. The evolving IGBP shares that property: its whole is more than the sum of its parts.

## Epilogue

To bring this story up to date, biogeochemical work has now been honoured in the Nobel Prize awards. The 1995 Chemistry Prize was awarded to Paul Crutzen, Mario Molina and Sherwood Rowland for their work on atmospheric chemistry and ozone. As stated by the Royal Swedish Academy of Sciences, "the connection demonstrated between soil microorganisms and the thickness of the stratospheric ozone layer is one of the motives for the recent rapid development of research on global biogeochemical cycles."

## Selected References

Broecker, W. S. and Denton, G. H. (1989) The role of ocean–atmosphere reorganizations in glacial cycles. *Geochimica et Cosmochimica Acta*, **53**, 2465–2501.

Butcher, S. S., Charlson, R. J., Orians, G. H. and Wolfe, G. V. (1992) *Global Biogeochemical Cycles*. Academic Press, London. 379 pp.

Charlson, R. J. and Wigley, T. M. L. (1994) Sulfate aerosol and climatic change. *Scientific American*, **270**(2), 28–35.

Dooge, J. C. I., Goodman, G. T., la Rivière, J. W. M., Marton-Lefèvre, J., O'Riordan, T. and Praderie, F. (1992) *An Agenda of Science for Environment and Development into the 21st Century*. Cambridge University Press, Cambridge. 331 pp.

Fajer, E. D. and Bazzaz, F. A. (1992) Is carbon dioxide a "good" greenhouse gas? Effects of increasing carbon dioxide on ecological systems. *Global Environmental Change*, **2**, 301–310.

International Geosphere–Biosphere Programme (1994) *IGBP in Action: Work Plan, 1994–1998*. Report No. 28, IGBP Stockholm, 151 pp.

International Geosphere–Biosphere Programme (1994) *IGBP Global Modelling and Data Activities, 1994–1998*. Report No. 30, IGBP Stockholm, 87 pp.

IPCC (1994) *Climate Change 1994: Radiative Forcing of Climate Change*, IPCC/Cambridge University Press, 338 pp.

Kuemmel, B. (1996) Editorial essay, *Climatic Change*, **32**, 379–385.

Lorius, C. and Oeschger, H. (1994) Palaeoperspectives: reducing uncertainties in global change? *Ambio*, **23, 30–36.**

Lovelock, J. E. (1991) *Gaia. The Practical Science of Planetary Medicine*. Gaia Books Ltd, London. 192 pp.

Martin, J. H. *et al.* (1994) Testing the iron hypothesis in ecosystems of the equatorial Pacific Ocean. *Nature*, **371**, 123–129.

McCormick, M. P., Thomason L. W. and Trepte, C. R. (1995) Atmospheric effects of the Mt Pinatubo eruption. *Nature*, **373**, 399–404.

Ojima, D. (Ed.) (1992) *Modelling the Earth System*. UCAR/Office for interdisciplinary Earth Studies, Boulder, USA. 488 pp.

Prinn, R. G. (Ed.) (1994) *Global Atmospheric–Biospheric Chemistry*. Proceedings of the first IGAC Scientific Conference. Plenum Press, New York. 261 pp.

Rind, D., Rozenzweig, C. and Goldberg, R. (1992) Modelling the hydrological cycle in assessments of climate change. *Nature*, **358**, 119–122.

Sahagian, D. L., Schwartz, F. W. and Jacobs, D. K. (1994) Direct anthropogenic contributions to sea level rise in the twentieth century. *Nature*, **367**, 54–57.

Schlesinger, W. H. (1991) *Biogeochemistry. An Analysis of Global Change*. Academic Press, San Diego. 443 pp.

Solomon, A. M. and Shugart, H. H. (Eds) (1993) *Vegetation Dynamics and Global Change*. Chapman & Hall, New York. 338 pp.

Sundquist, E. T. (1993) The global carbon dioxide budget. *Science*, **259**, 934–941.

Taylor, K. E. and Penner, J. E. (1994) Response of the climate system to atmospheric aerosols and greenhouse gases. *Nature*, **369**, 734–737.

Williamson, P. (Ed.) (1992) *Global Change: Reducing Uncertainties*. IGBP, Stockholm, 40 pp.

# 4. Understanding the Climate System

G. A. McBEAN
*Atmospheric Environment Service, Downsview, Ont., Canada (former Chairman ICSU-WMO-IOC Joint Scientific Committee, World Climate Research Programme)*

| | | |
|---|---|---|
| 1896 | S. Arrhenius, Swedish Nobel laureate | Suggested that emissions from burning of fossil fuels may change the climate |
| 1967 | WMO/ICSU Global Atmospheric Research Programme | Second objective – understanding the physical basis of climate |
| 1970 | Study of Man's Impact on Climate, Wijk, Sweden | Brought together leading experts to write a book on man's impact on climate |
| 1972 | Stockholm UN Conference on the Human Environment | Climate discussed, but not a major issue |
| 1979 | First World Climate Conference, Geneva | Primarily scientific and technical; led to World Climate Programme |
| 1980 | World Climate Programme, started by WMO, ICSU and UNEP | Included World Climate Research Programme |
| 1985 | International Conference on the Assessment of the Role of Carbon Dioxide and other Greenhouse Gases in Climate Variations and Associated Impacts, Villach, Austria | Conference statement called for governments to consider climate effects of greenhouse gases in policies on development, emissions and the environment |
| 1988 | Conference on the Changing Atmosphere: Implications for Global Security, Toronto | Called for 'Action Plan for the Protection of the Atmosphere': brought issues to world-wide government attention |
| 1988 | United Nations General Assembly resolution on Protection of Global Climate for Present and Future Generations of Mankind | Requested WMO and UNEP to take action on climate change issues |

*R. E. Munn, J. W. M. la Rivière and N. van Lookeren Campagne (eds), Policy Making in an Era of Global Environmental Change, 57–95. © Kluwer Academic Publishers. Printed in Great Britain.*

| | | |
|---|---|---|
| 1988 | WMO-UNEP Intergovernmental Panel on Climate Change | Mandate to assess scientific knowledge, environmental and socio-economic impacts and possible responsible response strategies |
| 1989 | Noordwijk Declaration on Climate | Set targets for emission reductions |
| 1990 | Bergen Declaration on Climate Change | Established the Precautionary Principle |
| 1990 | Second World Climate Conference and first report of the Intergovernmental Panel on Climate Change | Attended by Heads of State and Ministers from 137 countries; ministerial Declaration noted potentially severe impacts and called for global response |
| 1992 | Framework Convention on Climate Change (FCCC) | Signed during Rio Conference on Environment and Development; started process towards Climate Convention |
| 1994 | Conference of Parties – FCCC | First meeting, Berlin |

## Introduction – the Climate System

Climate is the statistics of day-to-day weather, mean values and fluctuations, for a particular region and time period. Although climate has been traditionally considered in terms of 30-year norms (e.g., the 1951–80 period), it is important to recognize that climate is dynamic, not static, varying on a multitude of time scales. The climate system involves the atmosphere, ocean, ice and land surfaces (including vegetation) which function as an integral whole (Figure 1).

The Earth's atmosphere is dominantly nitrogen (78% by volume), oxygen (21%) and argon (1%), plus water vapour ($H_2O$, up to a few per cent), carbon dioxide ($CO_2$, about 0.035%) and other trace gases with much smaller abundances but which play a major role in atmospheric dynamics and chemistry. Atmospheric concentrations of many of these trace gases are increasing, but some, such as stratospheric ozone ($O_3$), are decreasing. Solar radiation provides the energy that allows the biosphere to provide not only the fuel for life but the oxygen as well, through the process of photosynthesis that consumes $CO_2$ and releases $O_2$.

The oceans are the most massive part of the climate system. With an average depth of about 4 kilometres, the oceans cover 70% of the globe and provide vast storage of heat, water and trace constituents. Wind forces on the ocean surface create large horizontal gyres of ocean circulation, parts of which are the Gulf Stream and the Kuroshio Current.

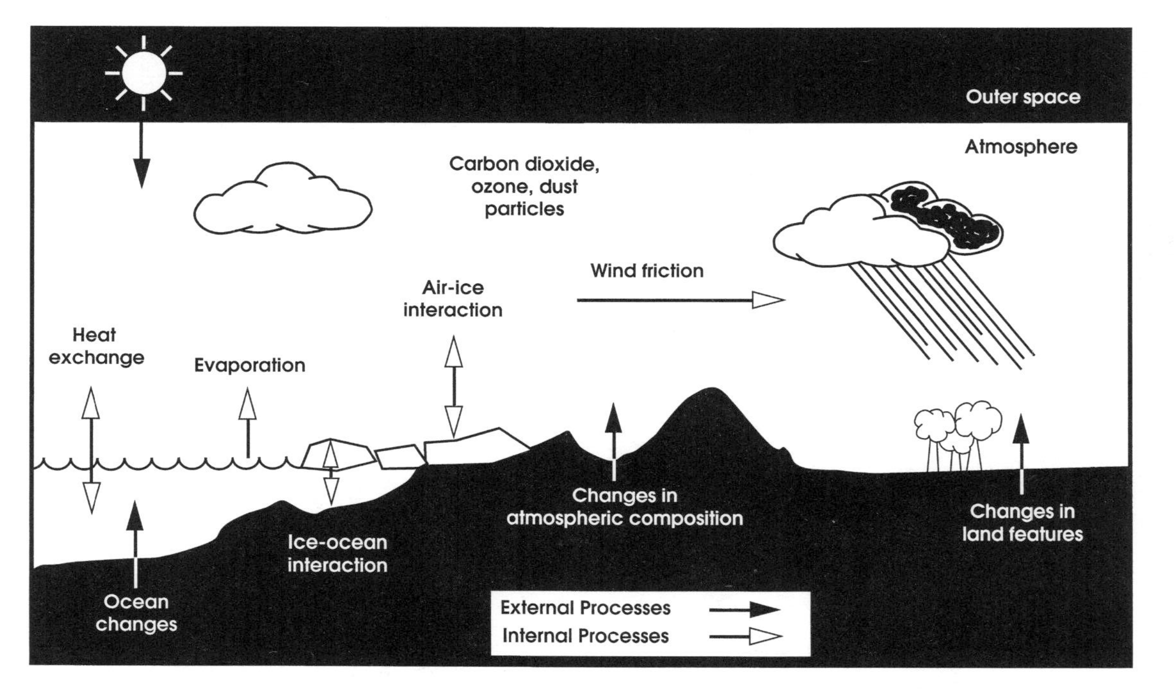


Fig. 1. The climate system.

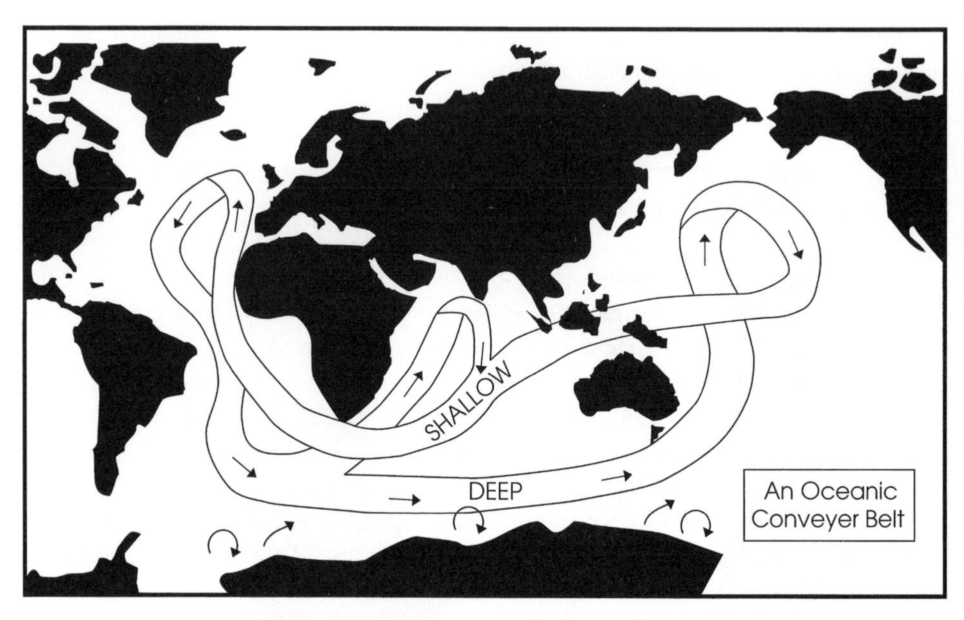


Fig. 2 The great ocean conveyor belt; a schematic representation of a major ocean current system linking the global ocean, that responds on time scales of years to centuries. Cold, salty water sinks in the North Atlantic Ocean and travels south as a deep current. The Antarctic Circumpolar Current connects the ocean basins. Cold and salty deep currents move northward in the Indian and Pacific Oceans, gradually mixing upward. The return current of warm water returns to the North Atlantic Ocean both around Africa and south of South America (not shown on schematic).

Transports up to 100 million cubic metres of water per second are common in these currents. Superimposed on these wind-driven horizontal gyres is an overturning current system, sometimes called the ocean conveyor belt (Figure 2), resulting from water sinking around Antarctica and in the northern North Atlantic. Without ocean currents and atmospheric winds transporting heat poleward, the Earth's equatorial regions would be much warmer than at present, and the poles much colder. The ocean conveyor belt plays a dominant role in the transport of heat and its strength depends on precipitation, evaporation and melting or freezing of sea ice at high latitudes.

Although large continental ice sheets (such as Greenland and Antarctica) are critical to climate evolution over thousands of years, sea ice and snow cover are more important over seasons to decades. Snow and ice reflect large amounts of solar radiation so their presence cools the climate. Winter snow cover on land surfaces also stores water, which is released several months later in the spring melt. Soils also retain moisture which can affect the next season's climate. Desert surfaces reflect twice the solar radiation than do grassy surfaces. Vegetation is also important in controlling evaporation from soil. Overgrazing in semi-desert areas, changing the radiative and evaporative properties, is thought to be a cause of desertification.

Vegetation on land and in the ocean also plays a major role in controlling the cycles of carbon dioxide, methane and other important gases that influence the climate system (see Chapter 3).

## The Natural Greenhouse Effect

The Earth's average temperature is determined by solar radiation intensity, planetary albedo (the fraction of solar radiation reflected back to space off clouds, aerosols and snow, ice and other surfaces) and the 'greenhouse effect' (see Box). The annual average solar radiation at the top of the atmosphere is 342 watts per square metre and the planetary albedo is 0.3 (i.e., 30% of incoming solar radiation is reflected back to space and the remaining 70% is absorbed at the surface or in the atmosphere). The amount absorbed by the climate system (about 240 watts per square metre) is approximately 30 000 times larger than the total world energy consumption through burning of oil, natural gas and coal and electricity generation by hydro and nuclear plants. The amounts of energy from geothermal sources and tidal power are minuscule by comparison.

Fortunately, the natural greenhouse warming effect takes average global temperature from well below freezing to well above; this is critical for the maintenance of life. It is interesting to compare Earth with its neighbouring planets. The atmosphere of Venus is dominated by carbon dioxide and has a very large greenhouse effect, raising Venus'

**The Natural Greenhouse Effect**

Every body emits energy (electromagnetic radiation), the amount depending on the surface temperature of the body. Hot things emit more energy than cold things, and the radiation from hot things has shorter wavelengths than the radiation from cold things. Because the sun is very hot (about 5700°C), it emits a tremendous amount of short-wave radiation (much of which is visible sunlight); the Earth is relatively cold and emits a small amount of long-wave radiation (invisible infrared radiation). This difference in radiation characteristics is very important in determining climate.

When skies are clear, solar short-wave radiation passes through the atmosphere with relatively little attenuation whereas long-wave radiation from the Earth's surface is largely absorbed in the atmosphere by molecules of the so-called greenhouse gases (water vapour (the most important one), carbon dioxide, methane, ozone and nitrous oxide), (see Figure 3). Because these greenhouse gases are good absorbers, they are also good emitters and, in turn, they re-emit energy both upward towards space and downward back to Earth. The colder molecules in the atmosphere emit less than the Earth's surface. The result is that the lower atmosphere and the Earth's surface are warmer than they would otherwise be. From space, the Earth appears to emit radiation at a temperature of −18°C (typical of air at about 5 km altitude). Greenhouse warming brings the surface temperature up to the observed 15°C, an increase of 33°C over the radiative temperature of −18°C.

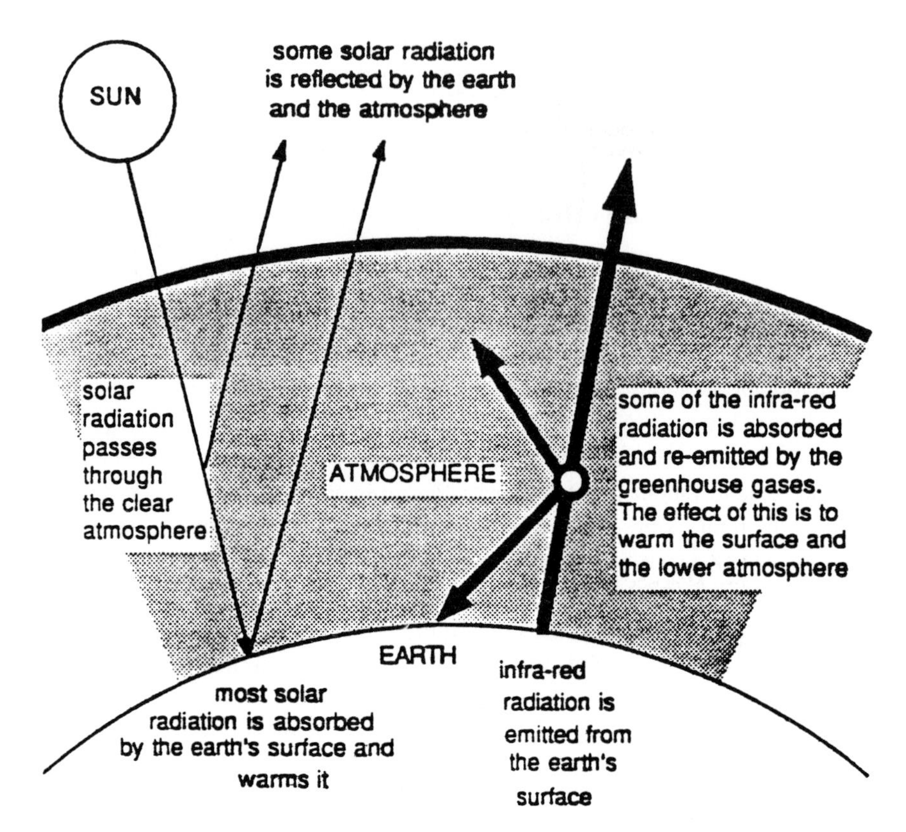


Fig. 3. A simplified diagram illustrating the greenhouse effect. Because the atmospheric greenhouse gases are colder than the Earth's surface, they emit less radiation to outer space than if there was no greenhouse effect.

surface temperature from –46°C to the observed 477°C. Mars, on the other hand, has very little greenhouse effect, 10°C, because of its thin atmosphere and the observed surface temperature is only –47°C.

## Temporal and Spatial Scales of Variability

The concept of scale is very important for scientific analysis. A football field is about 100 metres long and a very fast player could run its length in 10 seconds. These are its space and time scales. There is a wide range of scales in the climate system: a thunderstorm has a space scale of a few kilometres and a time scale of a few hours; a mid-latitude low-pressure system may cover 500 km and last for several days. For natural systems there is generally a correspondence between time and space scales: small space scale events have short time scales; large space scales have long time scales (Figure 4). When we deal with pollutants or constituents of the atmosphere, it is appropriate to speak of *residence time*, the

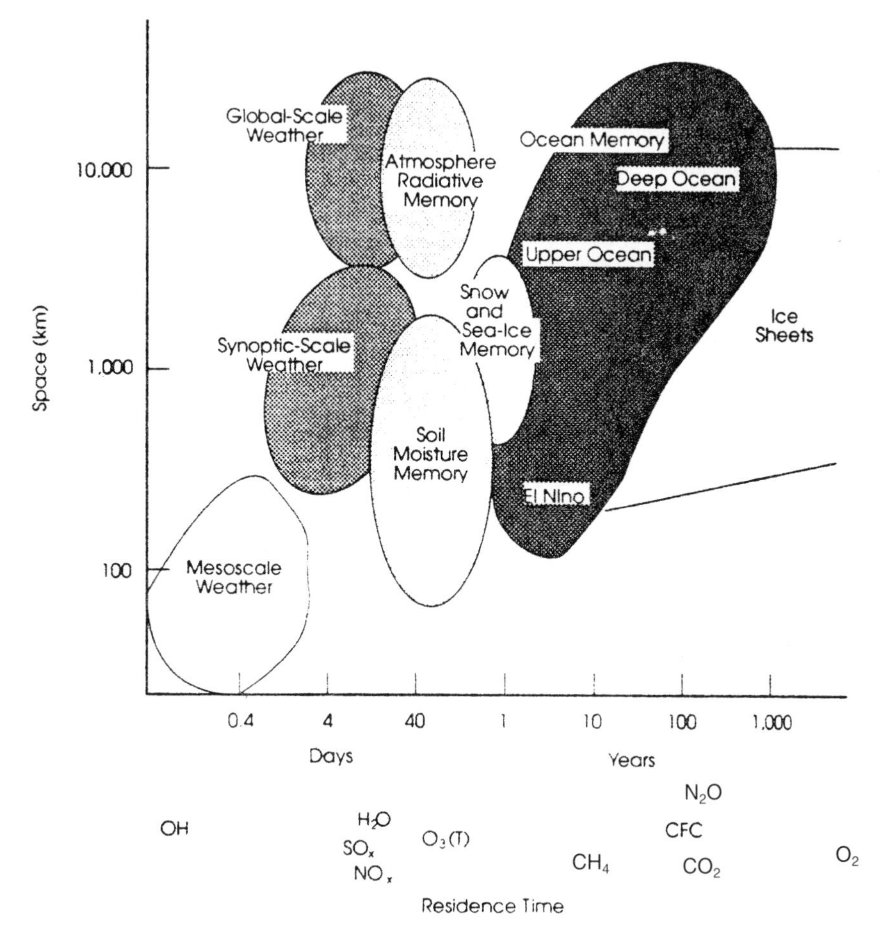


Fig. 4. Schematic representation of the time and space scales of the atmosphere and climate. Approximate atmospheric residence times of selected chemicals are indicated along the lower part.

typical time that a pollutant remains in the atmosphere. A similar concept is *adjustment time*, the speed with which the atmosphere adjusts to a significant change. Connected to this idea of time and space scales is the concept of predictability. It is not possible to predict the evolution of a weather phenomenon beyond a few multiples of its time scales.

Dynamically, the atmosphere adjusts in a few days; it will re-establish its radiative equilibrium in about a month. Water cycles through the atmosphere in about 10 days. The response time of the upper layer of the ocean (100–150 m thick) that responds directly to atmospheric forcing, is seasons up to a few years, while the response time of the deep ocean, which is beyond local atmospheric influence, is decades to centuries. Residence times for major ice sheets (Antarctica and Greenland) are hundreds to thousands of years. Adjustment times are a few months for soil moisture and snow and sea ice.

For pollutants, we need to consider the interactions of the atmosphere's natural time and space scales and those of the pollutant (including its time scale for chemical transformation). A pollutant injected into the lower atmosphere, which is generally well mixed, is spread horizontally by winds and mixed upward. Because water-soluble or particulate pollutants are either rained out or fall out, they usually do not stay in the lower atmosphere for more than a few weeks. Atmospheric concentrations are then a balance between emissions and removal processes. If a pollutant is injected into or reaches the stratosphere (typically 10–15 km up), where there is no rain and little vertical mixing, its residence time will be greatly lengthened due to the inefficiency of the removal processes. For many chemical species, chemical transformation may result in products that are more (or less) hazardous and less efficiently removed.

Carbon dioxide gas does not undergo chemical reactions in most of the atmosphere and its residence time, which is determined by complex interactions with the biosphere and the oceans, ranges from 50 to 200 years. The lifetime for methane is primarily determined by chemical reactions in the atmosphere and is estimated to be about 12 years. Nitrous oxide has the longest life of the main greenhouse gases, about 120 years. Water vapour has an atmospheric lifetime of about 10 days, very much shorter than any of the other important greenhouse gases. From a policy point of view, it is important to recognize that while remedial actions for short-lived constituents will result in quick responses, the consequences of actions for long-lived species will not be seen for many years.

It is useful to think of the climate system as being divided into two components: the fast and slow climate systems. The fast climate system is the atmosphere, the upper ocean and the transient processes at land surfaces. The fast climate system is active, driven by the atmospheric engine, and comes to statistical equilibrium in a few years. The slow climate system consists mainly of the deeper ocean and the perennial land ice, with a response time of decades to centuries. The major interactions between the fast and slow climate systems take place in a limited number of areas where heat is transferred by the up- or downwelling of ocean water and at high latitudes where cold, dense water sinks to great depths, i.e., the ocean conveyor belt. To a first approximation, the magnitude of climate response to a change in forcing, such as increasing greenhouse gas concentrations or changes in solar radiation, is determined by the fast climate system; the decadal rate of change is determined by the slow climate system.

## History of Earth's Climate: Natural Variability and its Impacts

### *Paleoclimate*

We think that the Earth was formed about 4.5 billion years ago, and paleoclimatic data, extracted from sediments, ice cores and other sources, indicate that the first large-scale ice age was about 2.3 billion years ago. During the last 2 million years, ice ages and interglacial periods (lasting about 10 000 years), have occurred on average every 100 000 years. We are presently in an interglacial period. Global surface temperatures typically decrease by about 5°C during the ice ages, with mid-to-high latitude decreases as great as 10–15°C.

Wobbles and asymmetries in the Earth's orbit around the sun have provided the pacemaker for geologically recent climate change (see Figure 5, upper panel). Three long-term astronomical cycles subtly affect the amount and distribution of solar energy received at the planetary surface: variations in axial tilt (with a periodicity of around 41 000 years); in the eccentricity of the Earth's orbit (100 000 years); and in their interplay – whether tilt effects are enhanced or weakened by distance effects, the 'precession of the equinoxes' (23 000 years). All three oscillations – the Milankovitch cycles – are evident in an 800 000-year record of fluctuations in the oxygen isotope ratios of marine sediments, an index of global ice volume, and can be traced in considerably more detail through the analyses of polar ice cores, covering the past 220 000 years.

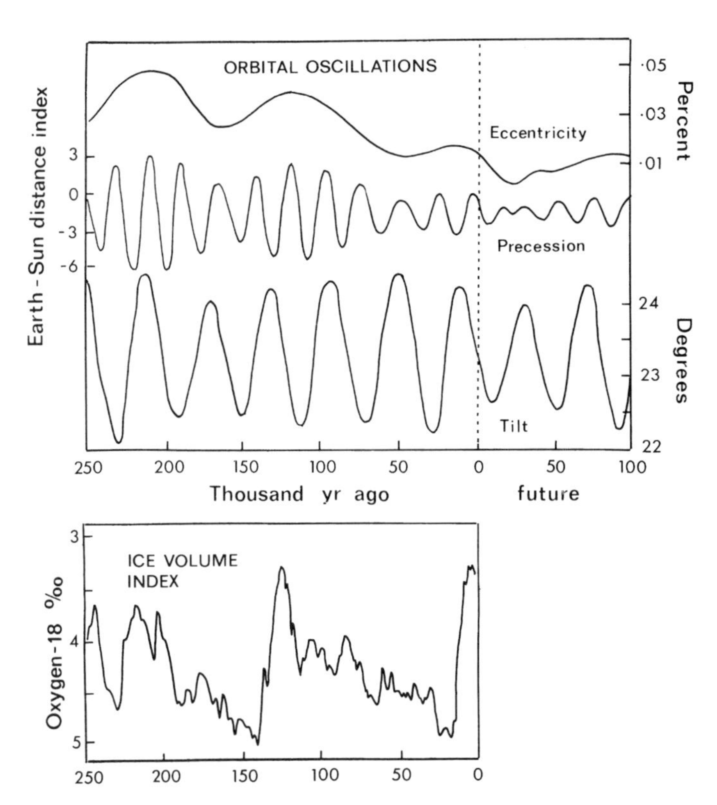


Fig. 5. The three main components of long-term changes in the earth's orbital geometry and their relationship to the ice-age cycle. Ice-volume data are based on oxygen isotope analyses from marine sediments. After N. G. Pisias and J. Imbrie 1986 (*Oceanus*, **29**, 43–49).

Paradoxically, the 100 000-year cycle is dominant in the climatic record (cf. upper and lower panels of Figure 5), although this cycle has the weakest effect on incident solar radiation, altering it by around 0.1% on a global basis. Warm, interglacial conditions (as currently experienced) are exceptional, and typically only last a few thousand years. A slow, cooling trend follows, and, with some fluctuations, polar ice cover grows. Over a period of nearly 100 000 years, surface temperatures fall by a global mean of 4–5°C; then a phase switch occurs, and glacial conditions rapidly wane. The other two cycles are superimposed on this saw-tooth pattern, causing secondary variations in global temperature of 0.5–1°C.

The main ice age oscillations are in phase with significant variations in incident radiation for the Northern Hemisphere. For example, at 60°N there was an increase in summer sunshine of around 10% (by nearly 40 watts per $m^2$) between 22 000 and 11 000 years ago, matching the end of the last ice age. Yet the Southern Hemisphere, where polar ice and mountain glaciers retreated at the same time, experienced a *decrease* in summer radiation, also of around 10% at 60°S. Because of these counteracting effects, global

**Rapid Climate Changes at the Dawn of Agriculture**

The end of the last ice age was followed by a marked transition in the development of human society – with many communities around the world changing from a nomadic lifestyle, as hunter-gatherers, to a more settled existence, based on agriculture. However, climatic stability is necessary for that transition, village and town populations becoming increasingly dependent on the successful harvests of their local farmers. Early attempts at agriculture seem likely to have been frustrated during the Younger Dryas period around 12 000 years ago, when, for a few hundred years, conditions of extreme cold returned to the Northern Hemisphere. The abrupt start, and end, of that event is clearly recorded in Greenland ice cores. Although direct evidence of the consequences for mankind is lacking, the scale of environmental disruption is shown by the dramatic changes in European vegetation (with *Dryas octopetela*, Mountain Avens, giving the name to the period); in the marine fauna of the northeast Atlantic; in the chemistry of Caribbean corals; and in the water levels of East African lakes (Figure 6). Taken together, these effects indicate that the most likely cause for the Younger Dryas event was a partial shut-down of deep water formation in the North Atlantic (brought about by meltwater), and that associated ocean circulation changes then radically altered regional temperature and rainfall patterns. The warming at the end of the Younger Dryas period was very rapid: the marine data indicate increases in sea surface temperatures of at least 5°C in less than 40 years, closely matching the ice-core estimates of a 7°C rise in South Greenland in about 50 years.

By 11 000 years ago, the more favourable conditions that characterize an interglacial period had fully returned. Since then, the Earth's climate system has been remarkably well-behaved. Is it merely a coincidence that the first large town supported by local agriculture (Jericho, in the Jordan valley) became established around 8500 BC, i.e. 10 500 years ago?

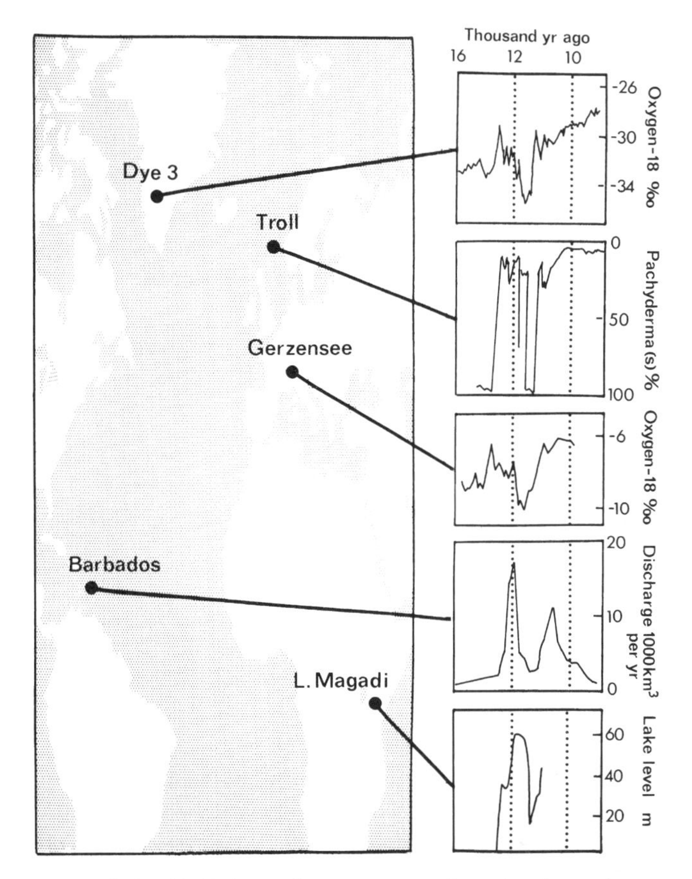


Fig. 6. Rapid climate switches 10–12 thousand years ago – as shown by changes in temperature-sensitive oxygen isotope ratios (at Dye 3 in Greenland and Gerzensee, Switzerland) in marine foraminifera (at the Troll coral site, off Norway). In the estimated Carribean meltwater discharge (from coral data at Barbados), and in African lake-water levels (for Lake Magadi, Kenya). After S. J. Lehman and L. D. Keigwin, 1992 (*Nature*, **356**, 757–762) and N. Roberts *et al.*, 1993 (*Nature*, **366**, 146–148).

climate models are unable to simulate the 100 000-year ice-age cycle solely from changes in solar energy. Several reinforcing mechanisms, both extraterrestrial and terrestrial, have therefore been proposed. These include changes in orbital inclination, with the same 100 000 year periodicity (with effects mediated via dust in space) and various inter-hemisphere linkages, involving switches in ocean circulation (affecting the deep-water conveyor belt), effects of land ice cover on sea level, land-mass elevation and continental bed-rock structure, and changes in atmospheric composition. Analysis of air bubbles that were trapped in the ice when it froze shows that atmospheric concentrations of greenhouse gases (carbon dioxide and methane) were generally 1.5 or 2 times greater during warm periods than during cold periods. At the time of maximum glaciation, ice sheets covered Canada and large areas of northern Europe and Asia; sea level was 120 m lower. The ice

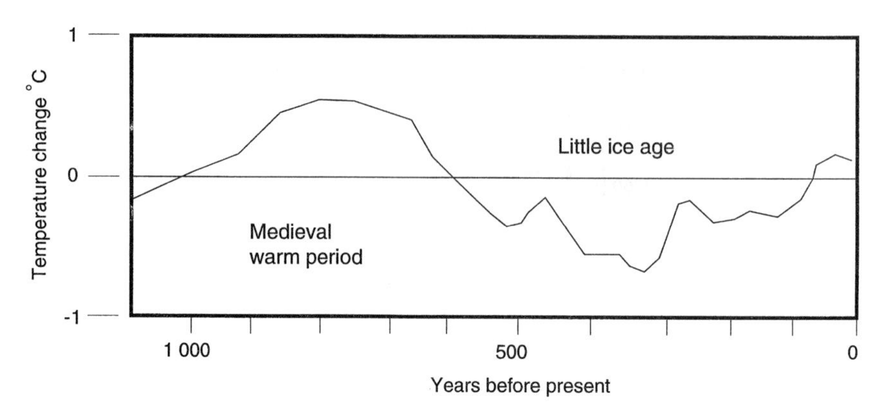


Fig. 7. Estimated variation in global mean temperature, relative to present climate, for the last 1000 years

ages must have caused great hardship for the populations of the time. However, the lowering of the seas allowed for migrations between continents and regions.

### *The Last Thousand Years*

Moving forward to the present millennium (see Figure 7), temperatures around the North Atlantic were exceptionally warm from the late 10th to the early 13th century when, for example, colonization took place in Iceland, Greenland and Labrador. But the most significant feature in the last 1000 years is the Little Ice Age, which occurred between the 14th and the 17th centuries when the settlements in Greenland perished, due to the failure of agriculture. Many events of European history were influenced by the anomalous weather conditions of the 15th through the 18th centuries. Invasions by the sea into Holland drowned hundreds of thousands of people in the 14th and 15th centuries. Floods and other natural tragedies have been common in Asia as well. An estimated 300 000 people drowned in China due to floods of the Hwang Ho River in 1642. In 1887, floods on the same river may have killed over a million persons and devastating floods have continued to occur through this century. A more subtle aspect of the interplay between large-scale climate anomalies and human society is shown in Figure 8 by the match between British temperature anomalies between 1660 and 1980, and switches in the local economic importance of herring and sardine fisheries.

### *The Last Century*

The Earth's climate has warmed 0.3–0.6°C during the past century, and the last decade has been particularly warm (Figure 9). The warming over the century has been mainly due to rapid increases from 1910 to 1944 and after the mid-1970s (both about 0.1°C per

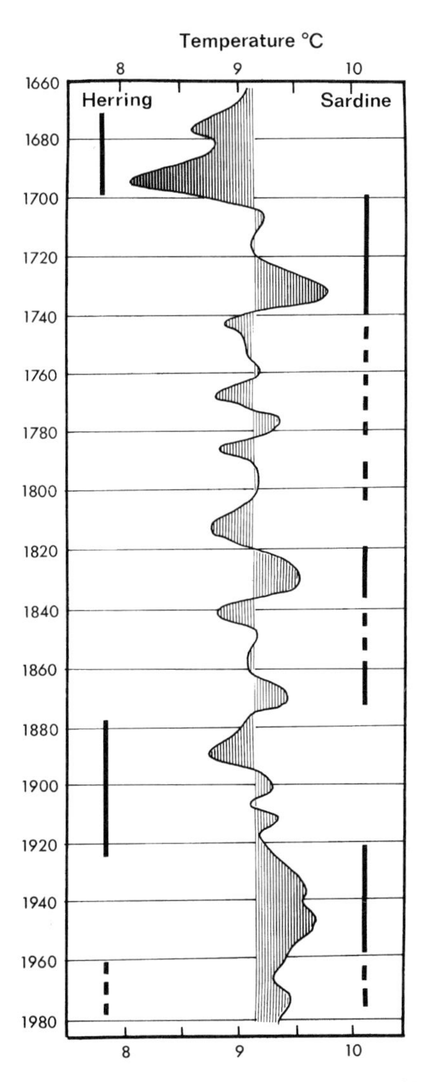


Fig. 8. A 320-year record of regional temperature fluctuations in central England, and the changes in the relative importance of the herring and sardine fisheries in Cornwall and Devon. Temperature data are an 11-year running mean, based on at least 25 records (pre-1720 data considered less reliable). The biological data are based on fishery statistics and other historical records. After J. Southward, G. T. Boalch and L. Maddock, 1988 (*J. Mar. Biol. Assoc. UK*, **68**, 423–445).

decade or 1°C per century). Sea level has risen 10–25 cm over the same period, mainly due to thermal expansion of the ocean and melting of small land glaciers. Although it is not known whether climatic variability or the frequency of extreme events has changed on the global scale, there have been significant increases in loss of life and economic damage in certain regions. For example, drought and famine in India during 1965–67 took some 1 500 000 lives. The prolonged drought in the Sahel of Africa through the

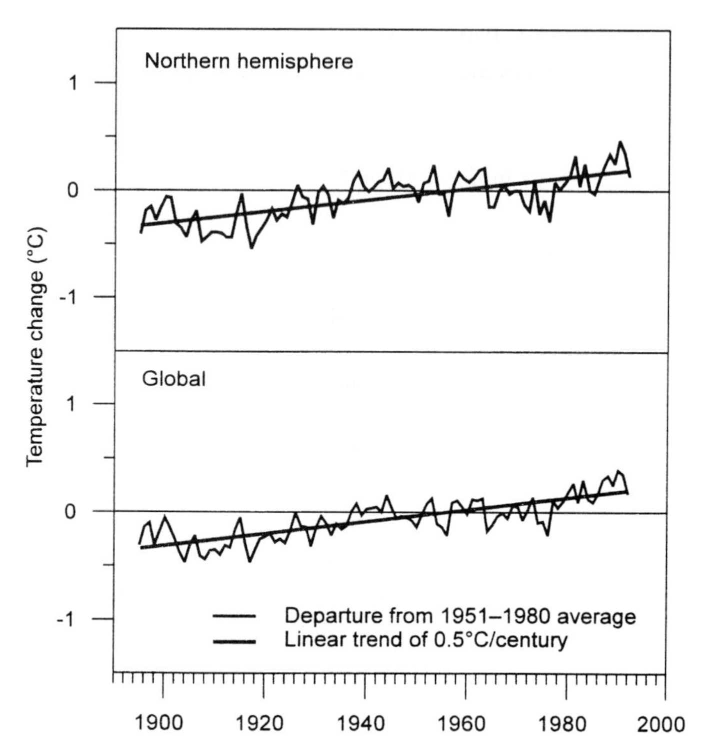


Fig. 9. Average surface annual temperature trends for the years 1895 to 1992 averaged for the Northern Hemisphere (upper plate) and globally (lower plate)

1970s and early 1980s killed many people, displaced many more and disrupted the lives of millions. Today, 36 African countries are affected by drought and 19 of those are among the least developed nations of the world. Drought has had a profound effect on the socioeconomic activities of these countries, including migration of pastoralists to the urban centres as well as triggering reductions in crop yields. In Africa, climate exerts significant control over development and further development needs to be planned carefully bearing in mind the sensitivity of the region to climatic variations.

Our understanding of the causes of these meteorological disasters has increased greatly in recent years. Coupling of the tropical Pacific Ocean and the overlying atmosphere, called the El Niño-Southern Oscillation (ENSO), results in anomalously warm sea surface temperatures in the eastern equatorial Pacific Ocean, on average every 4–5 years, which have global significance. In 1982–83 the strongest El Niño of the century occurred with extreme droughts in Australia, Indonesia, Bolivia and southern Peru and floods in northern Peru, Ecuador and Colombia. Through the winter and spring of 1982–83, rainfall in parts of northern Peru, Ecuador and Colombia was between twice and several hundred times normal amounts. Estimates of the damage due to the 1982–83 climate changes in Ecuador, Peru and Bolivia totalled over $US3 billion (over half in Peru). Less intense El Niño events also occurred in 1987 and 1991–93.

## Human Intervention – The Enhanced Greenhouse Effect

Although water vapour is the most important greenhouse gas, it is short-lived in the atmosphere and human activities do not seem to be changing its concentration. Carbon dioxide is and is expected to continue to be the dominant long-lived greenhouse gas for which human activities are having a major impact. The global cycle of carbon involves very large natural exchanges between the atmosphere and the ocean and land vegetation (see Chapter 3). Annually, land vegetation growth absorbs about 100 billion tonnes of carbon, in the form of carbon dioxide, which is returned to the atmosphere through vegetation decay and emissions from soils. Annually, about 100 billion tonnes of carbon are transferred, in both directions, between the ocean and the atmosphere. Generally, cold ocean waters absorb more carbon than they emit while warm ocean waters emit more than they absorb. In the cold ocean, biological and physical processes transport carbon to depth while other processes (e.g., upwelling) bring it upward in the tropics. These huge carbon cycles are not well understood nor quantified.

As human populations expand globally, their influence greatly increases. Initially, the impacts were felt locally, in cities and densely-populated areas. We know from written records of occurrences of urban pollution during the Roman Empire and through the Middle Ages. Combustion of wood and coal, and later oil, for heating, cooking, transportation and industry led to increases in pollutants in the atmosphere. However, these changes were very small on the global scale. It was not until the industrial revolution, starting in the 18th century, that pollutant emissions started to rise. It is estimated that global annual emissions of $CO_2$ from fossil fuel burning and cement manufacture were about 0.1 billion tonnes of carbon per year in 1860 and increased about 4% per year until the First World War (Figure 10). The increase slowed but total emissions passed 1 billion tonnes per year during the 1920s. Between 1950 and 1970, the rate of increase was again about 4% per year. The oil crisis of 1973 cut the rate of increase to half, to 2% per year, and emissions were nearly constant at 5.3 billion tonnes of carbon per year from 1979 to about 1985 before starting to increase again. Emissions of $CO_2$ are estimated to have been 5.7 billion tonnes carbon per year in 1987, with an uncertainty of 0.5 billion tonnes per year. For the period 1850 to 1987, it is estimated that 200 billion tonnes of carbon were released into the atmosphere from fossil fuel burning and cement manufacture.

The other main anthropogenic source of $CO_2$ is land-use change, including deforestation. Since pre-agricultural times, the world's forests are estimated to have decreased from about 5 billion hectares to 4 billion hectares. In the 1800s and during the first part of this century, carbon emissions due to deforestation exceeded those due to fossil fuel burning. With the growth of populations and pressures for development in developing coun-

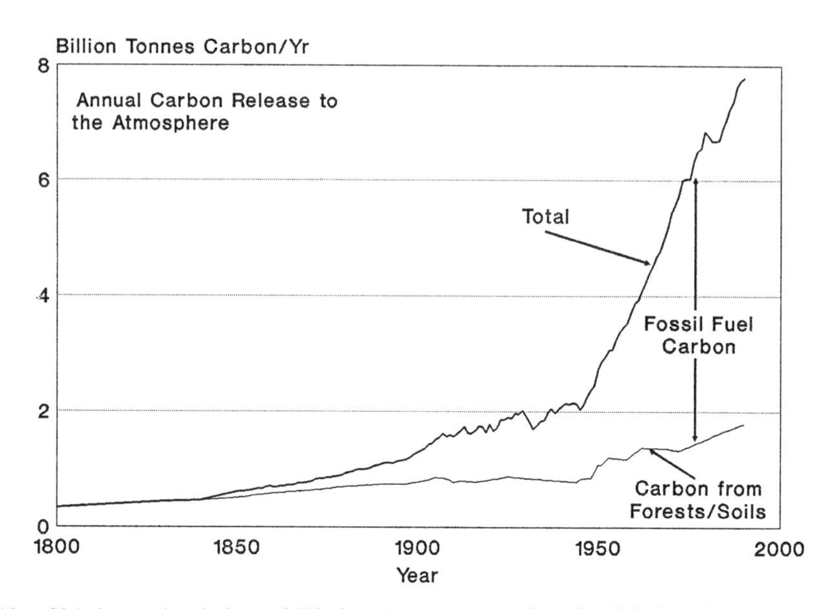


Fig. 10. Global annual emissions of $CO_2$ into the atmosphere from fossil fuel combustion and cement manufacturing and land-use change and deforestation and emissions from soils, expressed in billions of tonnes of carbon per year. The average rate of increase in emissions from fossil fuel combustion and cement manufacturing between 1860 and 1910 and between 1950 and 1970 is about 4% per year.

tries, percentage rates of deforestation have increased and are presently largest in South America (1.3% per year) and Asia (0.9% per year). Between 1850 and 1985, the total release of carbon to the atmosphere from land-use change, primarily deforestation, was about 115 billion tonnes. It is estimated that tropical deforestation has contributed about 2–3 times as much carbon as deforestation at mid-to-high latitudes.

Based on samples of air trapped in glacier ice, we know that the atmospheric concentration of carbon dioxide in the 18th century was about 280 parts per million by volume, ppmv (refer back to Figure 7, in Chapter 1, and see also Table 1). By 1910 it was 300 ppmv. By 1960 the atmospheric concentration was 320 ppmv; it reached 340 ppmv in 1980 and the rapid rise has continued; the 1992 global mean value was 355 ppmv (corresponding to 750 billion tonnes carbon). This is an increase of 28% over pre-industrial values. Now, the $CO_2$ concentration is increasing at a rate of about 3 billion tonnes carbon per year (0.4% per year, averaged over the 1980s).

The evidence is very strong that this increase is due to human activities. The atmospheric $CO_2$ concentration was relatively constant for at least 1000 years prior to the 19th century (within ± 10 ppmv of 280 ppmv). If land vegetation, oceans and atmosphere had not been in equilibrium, the atmospheric concentration would have been changing. From longer ice core data, going back 160 000 years, we can note that the maximum concentration was only about 300 ppmv during the last interglacial period 120 000 years ago. The increase in atmospheric carbon dioxide closely parallels estimates of accumulated emissions from land-use change and fossil fuel combustion.

Table 1. Pre-industrial and present concentrations, present emissions, lifetimes of greenhouse gases, ppmv, ppbv, pptv = parts per million, billion, trillion by volume

| | Carbon dioxide (ppmv) | Methane (ppmv) | Nitrous oxide (ppbv) | HCFC-22 (a CFC substitute) (pptv) | CFC-12 (pptv) |
|---|---|---|---|---|---|
| Pre-industrial concentration | 280 | 0.70 | 275 | 0 | 0 |
| 1994 concentration | 358 | 1.72 | 311 | 110 | 503 |
| Current rate of change/year | 1.5 | 0.013 | 0.75 | 7–8 | 18–20 |
| Percentage change/year | 0.4% | 0.8% | 0.25% | 7% | 4% |
| Lifetime (years) | 50–200 | 12.2 ± 3 | 120 | 12.1 ± 2.4 | 102 |

The 1994 IPCC Special Report concluded that present anthropogenic activities result in total emissions to the atmosphere of 7.1 ± 1.1 billion tonnes carbon per year, with deforestation and land-use change estimated to contribute 1.6 ± 1.0 billion tonnes carbon per year and fossil fuel burning, 5.5 ± 0.5 billion tonnes carbon per year. Since the atmospheric $CO_2$ concentration is increasing 3.2 billion tonnes carbon per year, 40–50% of the carbon emissions remain in the atmosphere. The remaining 4 billion tonnes are absorbed in the oceans or by the land biosphere, in approximately equal amounts (according to the best present estimates). It is important to understand how the world's ecosystems absorb this extra carbon because they may not be able to absorb carbon at this rate in the future.

Concentrations of other greenhouse gases have also increased. Methane was about 0.7 ppmv in pre-industrial times and is now 1.72 ppmv. This value is much higher than any other over the past 160 000 years. The rate of increase over the past few decades (0.8% per year) is almost twice that for carbon dioxide. Important sources of methane are natural wetlands, rice paddies and enteric fermentation in animals. Other human activities that contribute significantly to atmospheric methane are natural gas drilling, venting and transmission, biomass burning and landfills. An increase in soil moisture and temperatures at high latitudes could result in enhanced methane emissions. Most methane is removed from the atmosphere through chemical reactions with a naturally occurring substance, the hydroxyl radical. Due to increasing anthropogenic emissions, the oxidizing capacity of the atmosphere may be declining.

Values for nitrous oxide have increased from 275 ppbv (parts per billion by volume) in pre-industrial times to 311 ppbv now. The current rate of increase is lower than that for other greenhouse gases, 0.25% per year. Nitrous oxide is removed by photodissociation (breaking apart of molecules by the energy of solar radiation) in the stratosphere resulting in its long (120 year) lifetime. Sources of nitrous oxide are not well quantified but emissions from tropical soils are important. The human role seems mainly in combustion, biomass burning and use of fertilizers.

The greenhouse gases showing the most dramatic rate of increase are the chlorofluorocarbons (CFC) that human ingenuity only invented about half a century ago. The atmospheric concentration of CFC-12 is increasing at 4% per year. The CFCs are used as aerosol propellants, refrigerants, foam blowing agents, solvents and fire retardants and are mainly removed by photodissociation in the stratosphere, a process that leads to destruction of the ozone layer. Through the Montreal Protocol on Substances that Deplete the Ozone Layer and follow-on agreements, production of CFCs will be either eliminated or greatly reduced by 1997. However, the atmospheric concentrations of CFCs 11, 12 and 113 will still be 30–40% of their current values for at least the next century because of their long atmospheric lifetimes, which vary with species but the most prevalent ones have lifetimes of about a century (CFC-12, 102 years).

Depletion of the stratospheric ozone layer is a major environmental concern itself. However, changing ozone concentrations are also important for the radiative forcing of climate because ozone is a greenhouse gas. The observed global lower-stratospheric ozone depletions are calculated to have cooled the stratosphere at middle and high latitudes consistent with observations of cooling there. This cooling will cause a decrease in the radiative forcing of the troposphere. Ozone depletion over the past decade may have offset a significant fraction of the radiative forcing increase due to increases in greenhouse gases (UNEP-WMO Scientific Assessment of Stratospheric Ozone, 1991). Due to the role of CFCs in destroying ozone, their net effect on climate change is much smaller than earlier thought. We, thus, have the situation where the climatic influence of a pollutant with greenhouse gas properties (CFCs) may be offset by its influence on another greenhouse gas (ozone).

Due to differing radiative properties of molecules, changes in radiative forcing due to changes in the masses of greenhouse gases are not in direct proportion to mass changes. Relative to a unit mass of carbon dioxide, a unit mass change in methane is 56 times more effective in changing radiative forcing. For nitrous oxide the ratio is 280 times more effective, while for CFC-11 and CFC-12 the ratios are very large, between 2000 and 6000. When these relative radiative forcing factors and indirect effects were incorporated into the calculations, the relative contributions towards enhanced greenhouse radiative forcing for the decade of the 1980s were: carbon dioxide (70%); methane (20%); and nitrous oxide (5%) (all rounded to one significant figure); and CFCs (24%). This computation did not include the influence of ozone.

A common feature of these greenhouse gases is their relatively long lifetimes in the atmosphere; ranging from 12 years for methane to about a century for the others. A consequence is that the gases accumulate in the atmosphere. Today's atmosphere holds much of this century's carbon dioxide, nitrous oxide and CFCs emissions. Hence, as long as emissions exceed the capability of the oceans or terrestrial ecosystems to absorb the gas or for chemical transformation or other processes to remove it, atmospheric concentrations will increase, even if our emissions are not increasing. For the long-lived greenhouse gases, dramatic reductions are required in order to stabilize the atmospheric concentrations at present-day levels (Table 2).

Table 2. Reductions in anthropogenic emissions of greenhouse gases required to stabilize the atmospheric concentrations at present levels

| Greenhouse gas | Reduction required |
|---|---|
| Carbon dioxide | > 60% |
| Methane | 15–20% |
| Nitrous oxide | 70–80% |
| CFC-11 | 70–75% |
| CFC-12 | 75–85% |

Because any human-induced climate change will be superimposed on the background of natural climate variability, detection and attribution of the causes is difficult. However, observed global warming averaged over the past 100 years is larger than best estimates of natural variability over at least the last 600 years. Temporal and spatial variations (both horizontally and vertically) in the observed climate record are consistent with the emerging pattern of climate response to forcings by greenhouse gases and sulphate aerosols (see Fig. 11, p. 47). These results point towards a detectable human influence on global climate.

## Climate Research – International Coordination – National Efforts

Concern about climate change and global warming is based on climate research. Scientists, motivated by their scientific curiosity, have tried to understand how the climate system functions. They have delved into the processes by which certain gases absorb and emit radiation and have constructed complex models of the climate system. Scientists also started measurement programmes showing that the atmospheric concentration of $CO_2$ is increasing and developed the techniques to explore the atmospheric composition in the past, through the extraction of trapped air bubbles from ice cores. Much of this climate research has been conducted by individuals working with funding from national sources for individual research grants. However, climate is a global issue and internationally coordinated research programmes are needed. In 1980, the World Climate Research Programme was started by the World Meteorological Organization and the International Council of Scientific Unions; the Intergovernmental Oceanographic Commission (IOC) of Unesco joined as a partner in 1991. The World Climate Research Programme (WCRP) is a component of the World Climate Programme. The objectives and core programmes of the WCRP are given in Box 1. WCRP activities over the past decade have provided a much better understanding of climate variability and change and over the next decades will be instrumental in reducing our present uncertainties. Without internationally coordinated, large-scale programmes, it is unlikely that uncertainties in our understanding of the climate system will be substantially reduced.

| Box 1. The World Climate Research Programme |
|---|
| The objectives of the World Climate Research Programme are to determine: |
| to what extent climate can be predicted; and<br>the extent of human influence on climate. |
| The core programmes of the World Climate Research Programme are: |
| Tropical Ocean and Global Atmosphere Programme (TOGA)<br>Global Energy and Water Cycle Experiment (GEWEX)<br>World Ocean Circulation Experiment (WOCE)<br>Arctic Climate System Study (ACSyS)<br>Stratospheric Processes and their Role in Climate (SPARC)<br>Climate Variability and Predictability (CLIVAR) |

It must be emphasized that the WCRP is only a mechanism for international planning and coordination; the WCRP has no resources to carry out research. All of the WCRP programmes and projects are carried out by scientists and staff, ships, aircraft, computers and other facilities that are paid for by nations. Without the support of nations there would not be a WCRP, or any other international research programme. Discussions with national research institutions are an ongoing part of the WCRP development process. When the plans for a major project are completed (itself an interactive process involving scientists and research managers from many nations), nations are invited to implementation conferences where national contributions are matched with project elements. Only when the contributions are sufficient for a successful experiment, is the decision made to proceed. Through this process of programme planning and implementation, the WCRP relies heavily on its parent organizations, the World Meteorological Organization, the International Council of Scientific Unions and the Intergovernmental Oceanographic Commission, to provide contacts with national organizations and with the scientific community.

## Prospects for Climate Prediction

Although it is of considerable interest to be able to understand the climate system, the benefits of this understanding only really manifest themselves if we can turn that understanding into an ability to predict, with confidence, future climates. We all have experience with weather forecasts, with economic forecasts and with predictions on outcomes of elections and sporting events. Each involves essentially the use of some information which is put into a model, resulting in a prediction. The information may be voluminous, such as all the weather observations around the globe, or relatively little, such as the guesses of several panel members on election eve. The model also may be complex or, in its simplest sense, may be just the intuition of the predictor. The confidence that is placed in the prediction depends on the input information and the model, but most importantly on the past history of success or failure. We have little confidence in the fifth prediction if the first four have been wrong. Forecasters establish skill through the production and verification of predictions.

Weather forecasting has developed greatly over the past century. Presently, weather forecasts are based on large global observing systems, feeding information to the world's largest computers, which run very complex models, which in turn are used by well-trained weather forecasters to provide a valuable service to the public. Forecasts are released daily and are valid for periods of a few days to a week. The skill of the forecast system can be quantitatively measured every day and over the years high statistical confidence in the skill of the system can be gained. Through this procedure it has been possible to document the improvement in prediction skill over the last half century; very significant improvements have been made. Improvements in prediction skill have arisen due to improved input information (better observations and communications), better models (and faster computers to make them useful for real-time prediction) and more skill in the forecasters' interpretation of observations and models to give a better forecast.

The skill of weather forecasts is usually compared with a climatology forecast and a persistence forecast (Figure 11a). A climatology forecast is the average weather for that location and that time of the year. A persistence forecast is for weather to continue as it is. Clearly, if a weather forecast is no better than both climatology and persistence, it has no useful skill and cannot be justified economically. For very short time periods (a few hours) a persistence forecast will be accurate but, as the weather evolves, a persistence forecast will quickly become useless.

Weather forecasts are predictions from an initial state of the atmosphere to some time ahead. The atmosphere has a short memory and its prediction from an initial state is limited according to the spatial scale under consideration. The predictability for thunderstorms and other dangerous mesoscale events is less than a few hours. For major mid-latitude weather systems, the theoretical predictability limit is thought to be about two weeks. Computer models are limited by the spatial scales of the input data, by model resolution and by the model itself. The spacing between weather observations varies considerably around the globe, from very few observations over the oceans and parts of the developing world to relatively good observations for most of Europe and North America and parts of Asia. Weather prediction models now simulate the atmosphere on grids of about 100 km. Weather elements smaller than a few grid lengths cannot be represented in the model and hence cannot be predicted, except statistically. At present, weather forecasts prepared by major weather centres have skill up to about a week. For that period, their forecasts are more likely to be correct than just using the climatology for the day. The objective of forecast systems is to increase the level of skill for all periods and to extend the length of the forecast period that shows skill.

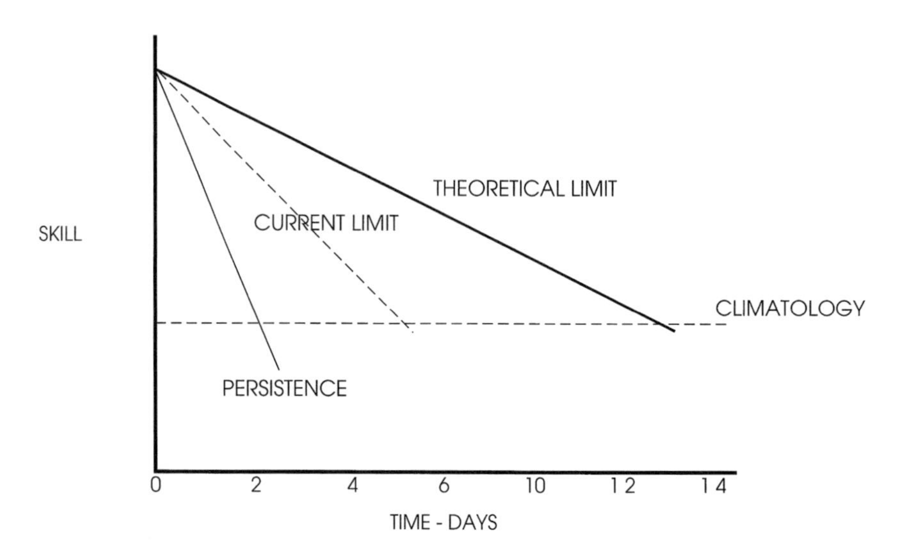
(a)
WEATHER PREDICTION FROM INITIAL STATE
SKILL
THEORETICAL LIMIT
CURRENT LIMIT
CLIMATOLOGY
PERSISTENCE
0
2
4
6
10
12
14
TIME - DAYS

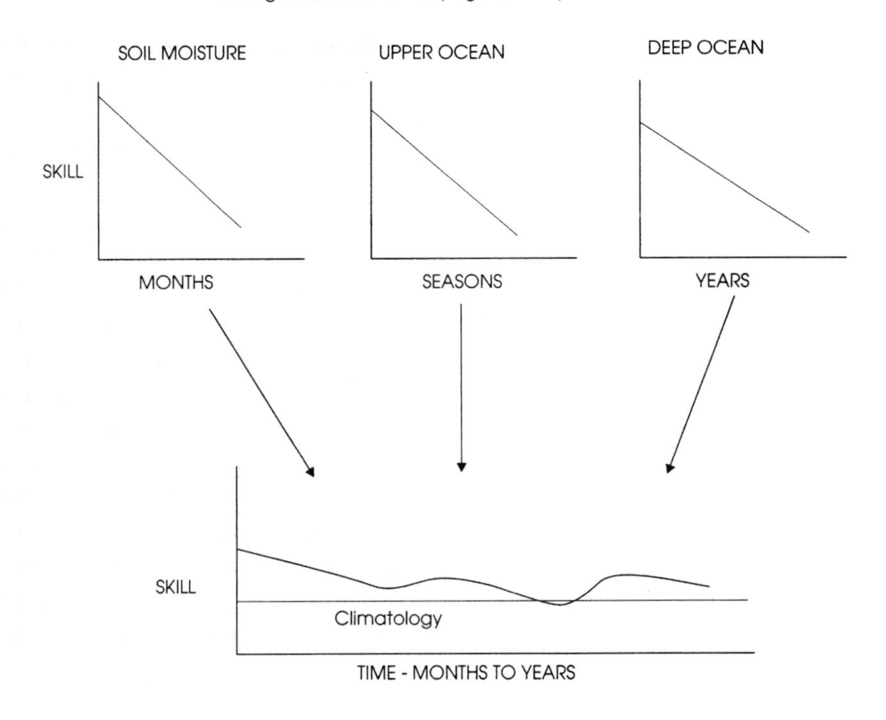
(b)
CLIMATE PREDICTION
Statistical Prediction of Climate Anomalies
Through Prediction of Varying Boundary Conditions
SOIL MOISTURE
UPPER OCEAN
DEEP OCEAN
SKILL
MONTHS
SEASONS
YEARS
SKILL
Climatology
TIME - MONTHS TO YEARS

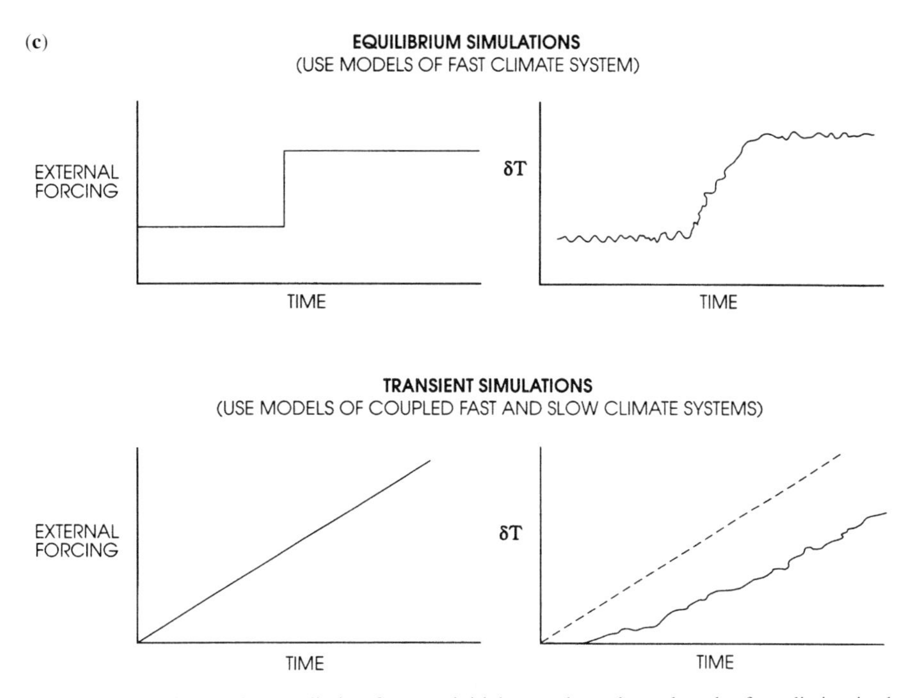


Fig. 11. **(a)** Skill in weather prediction from an initial state depends on length of prediction in days. Persistence forecasts are based on the assumption of no change from initial conditions, while the climatology forecast assumes the long-term mean for that day and location. Current forecast skill is schematically indicated by the long dashes and the theoretical limit by the solid line. **(b)** Skill in climate prediction from an initial state relies on the longer memory of soil moisture, upper ocean and deeper ocean conditions. **(c)** Climate scenarios are predictions of climate changes due to changing external forcing such as greenhouse gas concentration or solar output.

Beyond the limit of weather predictability, there is possibility of general predictions of longer time-scale features which depend on the longer time scales of the land surface and particularly, of the oceans (Figure 11b). For example, if a particular type of weather is related to upper ocean heat content, then forecasts may be made on the basis of ocean parameters. Skill in predicting the occurrence of rainfall anomalies, in a statistical sense, based on oceanic sea surface temperature anomalies has been demonstrated for periods of several months. As our understanding of these predictions improves, better monthly to seasonal forecasts will be possible. Climate predictions are the prediction of the statistical ensemble of weather events, what we call climate, for some period ahead based on observations of the initial state of the climate system (atmosphere, ocean, land surface) and use of a model of the coupled climate system.

Since the basis for these longer-period forecasts lies in the longer term memory of land surfaces and oceans, it is imperative that we observe, understand and model those components of the climate system, as well as the atmosphere. New satellite systems will provide improved global coverage of ocean surface characteristics (wind, temperature and topography), but we will still need measurements within the oceans. To provide the observational basis for prediction of natural climate variability, WMO, ICSU, IOC and UNEP are developing the Global Climate Observing System.

There are inherent limitations to predictions based on the initial state, due to inaccuracy and lack of spatial detail of observations, as well as the predictive skill of models. Although the oceans are slower to change than the atmosphere, eventually they too exceed their predictability limit. The next possibility of prediction then becomes prediction based on external conditions that are prescribed essentially outside the climate system (Figure 11c). To a large extent these are 'what-if' forecasts. For example, *what* will the climate be like *if* the amount of greenhouse gas in the atmosphere were to increase to double its present value and beyond; or, *what* will the climate be like *if* the Sun's intensity were to decrease? These are climate scenarios based on external conditions and are the basis of the present concern about global climate warming if greenhouse gas concentrations continue to increase.

It is important to recognize the three types of predictions: *weather predictions* from initial states; *climate predictions*, also from initial state of the climate system; and *climate scenarios* based on external conditions that are essentially beyond the climate system (such as human input of greenhouse gases). We have skill at the first type, which can be well documented. Our skill at climate predictions is generally unproven and is a new area of endeavour. Our skill at climate scenarios prediction is also relatively unproven. Some skill has been demonstrated in model simulations of past climates when the intensity of incoming radiation at the top of the atmosphere was different. However, our data for measurement of skill is limited. Our skill at climate scenarios due to the external condition of increasing concentrations of greenhouse gases will only be proven over the next several decades, which is too long to wait if we wish to take remedial action. However, if we develop our climate models on the same principles as weather prediction models and similar models for oceans and land surfaces (and can demonstrate skill for them), we should have some confidence that our models are useful (have some skill) for longer period predictions based on changing external conditions. In addition, climate models can be checked for their ability to model present climate with its time and space variations. Further, it can be noted that climate models do well at simulating the annual cycle; the difference between summer and winter at most latitudes is greater than the changes expected over the next few centuries for climate warming.

## The IPCC Scenarios for Future Climate

The evolution of the Earth's climate over the next century will result from ongoing natural variations combined with a warming trend forced by anthropogenic modifications of the atmospheric concentrations of greenhouse gases and aerosols. Any prediction of actual climate change depends on the rate of increase of atmospheric concentration of greenhouse gases, which depends on the emissions from human activities as well as natural processes. The Intergovernmental Panel on Climate Change (IPCC) prepared two scientific assessments of climate change and made projections on future climate change. These scenarios of future climate were based on scenarios of future emissions of greenhouse gases. Preparation of scenarios of future emissions depends on scenarios of population growth, energy use and a wide variety of other factors. For its first Assessment in 1990 the IPCC used the four scenarios described in Figure 12a and Table 3. In 1992 the IPCC developed a new set of scenarios and this work continues. The scenarios generally overlap in terms of the range of emissions, and only the 1990 scenarios will be used here.

Various other assumptions are included in these scenarios, which are presented for illustrative purposes and not as recommendations. For each scenario, there were also emission estimates for methane, nitrous oxide and CFCs. Actual anthropogenic emis-

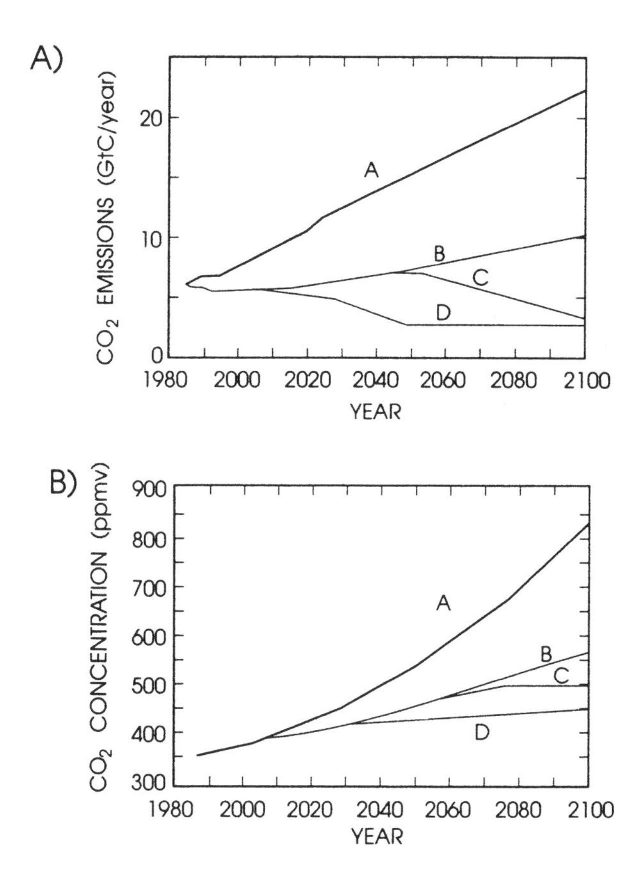


Fig. 12. **(a)** IPCC carbon dioxide emissions scenarios, in billions of tonnes carbon per year. **(b)** IPCC atmospheric carbon dioxide concentrations, in ppmv. See Table 3, p. 82, for meanings of letters A, B, C and D.

Table 3.

| Scenario | | Predicted rate of increase |
|---|---|---|
| **A** | Energy supply of the future is assumed to be coal-intensive and on the demand side with only modest efficiency increases achieved. Anthropogenic carbon dioxide emissions will increase approximately linearly from 7 billion tonnes carbon per year in 1990 to about 22 billion tonnes carbon per year in 2100. | About 0.3°C per decade (with an uncertainty range of 0.2°C to 0.5°C per decade) |
| **B** | Energy supply mix is shifted towards lower carbon fuels. After no increase until about 2010, the emissions increase linearly to about 10 billion tonnes carbon per year in 2100. | About 0.2°C per decade |
| **C** | A shift in energy supply to renewables and nuclear power takes place in the second half of the 21st century. Scenario C follows B until 2050 and then emissions decrease to 4 billion tonnes carbon per year in 2100. | Just above 0.1°C per decade |
| **D** | A shift to renewables and nuclear energy is assumed in the first half of the 21st century. The emissions change little until 2020, then decrease to about 3 billion tonnes carbon per year by 2050 and then remain constant until 2100 | About 0.1°C per decade |

sions of greenhouse gases to the atmosphere depend on social, economic and political factors that are difficult to predict. These scenarios did not include the implications of emissions of sulphate aerosols (see Figs. 11, 13 on pp. 47, 49).

Once emitted to the atmosphere, greenhouse gases come into the realm of study of the natural scientist. Presently, about 45% of the anthropogenic emissions of $CO_2$ remain in the atmosphere and the remainder are absorbed in the oceans or on land. For the IPCC scenarios, models based on our present understanding were used to estimate future atmospheric concentrations (Figure 12b). For Scenario A, atmospheric $CO_2$ doubles by mid-21st century. Because of the long lifetimes of greenhouse gases, the atmospheric concentration of carbon dioxide will increase through the next century, even for scenario D with its decrease in emissions to less than 50% of 1990 levels by 2050. It was assumed that natural cycles will not change. However, it should be noted that warming of the surface ocean will increase the area that is outgassing carbon dioxide and decrease the area of uptake. It will also reduce deep convection and further reduce the uptake of carbon dioxide by colder waters. Warming and melting of permafrost areas may also release large amounts of methane. There is potential for large positive feedback processes involving physical, chemical and biological processes in the biogeochemical cycles of the greenhouse gases that could lead to rapid increases in gas concentrations. Hence, it is possible that atmospheric concentrations could increase much more dramatically, for a given anthropogenic emission scenario, than indicated in Figure 12b.

Given the atmospheric concentrations, the radiative forcing due to the additional greenhouse gases can be computed using radiative transfer models. These depend on clouds, aerosols and other gases, as well as the thermal structure of the atmosphere. The time-dependent response of the climate system to this additional radiative forcing depends on complicated responses of both the fast and slow climate systems. Although the fast and slow systems are not independent, we can, to a first approximation, examine them sequentially. Adjustments in the fast climate system involve feedback processes which amplify or reduce the primary greenhouse effect. The slow climate system is controlled by the global ocean which sets the pace for climatic change and may introduce a delay of decades in the transient response of the Earth's climate to greenhouse forcing.

Doubling of atmospheric concentration of $CO_2$ results in a 4 watts per square metre (W $m^{-2}$) increase in the radiative forcing on the lower atmosphere. This is a small change compared to the typical fluxes of 50–100 W $m^{-2}$ in the atmosphere or the average solar irradiance at the top of the atmosphere (342 W $m^{-2}$). In fact, 4 W $m^{-2}$ is smaller than our present measurement accuracy. The climate system absorbs the 4 W $m^{-2}$ additional energy and warms up in order to increase the outgoing radiation and re-establish radiative equilibrium. If we look at the Earth from outer space, we will not see any difference. If no feedback or other complicating processes were involved, global mean surface temperature would warm by about 1°C. However, there are feedbacks in this non-linear system so the response is not straightforward.

There are at least three important feedback processes that change the climate system's response: water vapour and snow-ice albedo feedbacks (Figure 13) which are both positive and hence amplify the climate change; and cloud feedback (Figure 14), which may be positive or negative. Water vapour feedback is the most straightforward. If the temperature does increase due to extra radiative heating caused by increased $CO_2$ concentration, air can hold more water vapour and the warmer surface will evaporate more. Since water vapour is a greenhouse gas, increasing the concentration of one greenhouse gas ($CO_2$) causes an increase in another ($H_2O$), resulting in a positive or amplifying feedback mechanism. A second positive feedback mechanism is snow-ice albedo feedback. A warmer Earth with less snow will absorb more solar radiation.

Clouds play a major role in controlling heat input to the Earth's surface as well as the radiative heat loss to space (Figure 14). Low clouds generally cool the planet by reflecting more solar radiation back to space, although they also trap radiation below them. High cirrus clouds generally warm the planet since they reflect little solar radiation but trap Earth's radiation. Although the balance is very delicate, our best measurements from space indicate that the present distribution of cloudiness cools our planet. How clouds will influence climate change depends on feedback mechanisms which are extremely complex and depend upon: cloud amount; cloud altitude; and cloud water content. Reduced cloud amount decreases the greenhouse effect and acts as a negative feedback. However, solar radiation reaching the surface will increase – a positive feedback. If clouds move to higher altitude, they have a stronger greenhouse effect – a positive feedback. Changing the cloud water content, and depending on whether the cloud is solid or liquid, leads to very complicated interactions, with some models indicating positive and

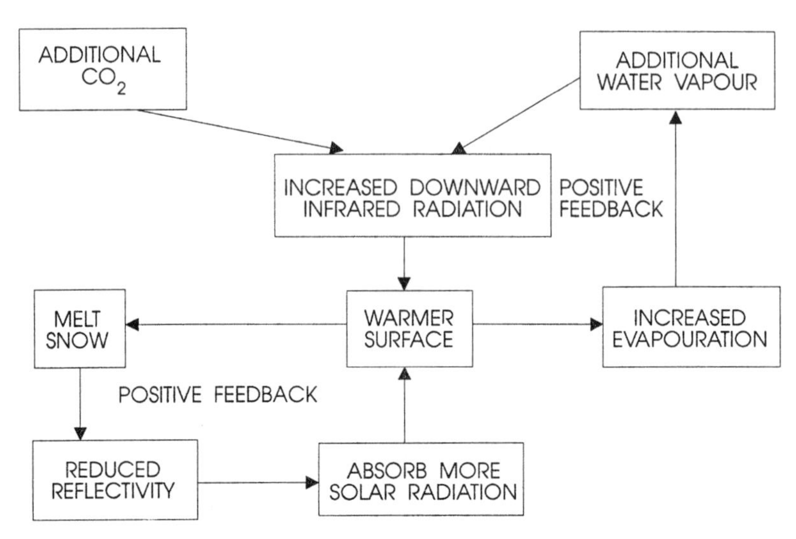


Fig. 13. Examples of positive feedbacks or amplifiers of the climate system that result in greater climate warming than just due to additional carbon dioxide greenhouse effect. Water vapour feedback results from additional evaporation resulting in more water vapour in the atmosphere which increases the greenhouse effect since water vapour is also a greenhouse gas. Snow-ice/reflectivity feedback results from melting of snow and ice (or changing surface temperature) which reduces the reflectivity of the surface causing it to absorb (rather than reflect) more solar radiation.

others, negative feedbacks. Present models indicate that in a greenhouse-gas warmed climate, the cooling due to clouds would decrease and hence cloud feedback would be positive – an amplifier of climate change. However, it is not possible to compute the strength of this feedback with confidence.

The standard scientific approach towards understanding these interacting feedbacks would be to conduct an experiment under controlled conditions, vary a set of parameters and observe what happens. Unfortunately, the climate system is too complex to reconstruct in a laboratory. Although it can be argued that we are inadvertently doing an experiment on the climate system through adding greenhouse gases, the experiment will take decades and we cannot undo its impacts. This experiment is not ethically, environmentally, or scientifically acceptable. The only way we can study climate system response is through use of very sophisticated global climate models.

Model experiments are initially conducted with a typical value of carbon dioxide before major human intervention. The models can then be compared with the climate of the Earth over the past few decades. Generally, climate models show considerable skill in the portrayal of the large-scale distribution of air pressure, temperature, wind and precipitation in both summer and winter. On regional scales, however, there are significant differences between models and observations. Because of computer limitations, climate models represent the Earth's surface in terms of 3 or 4 degree latitude–longitude squares so that one point represents the Netherlands–Belgium–Luxembourg, for example,

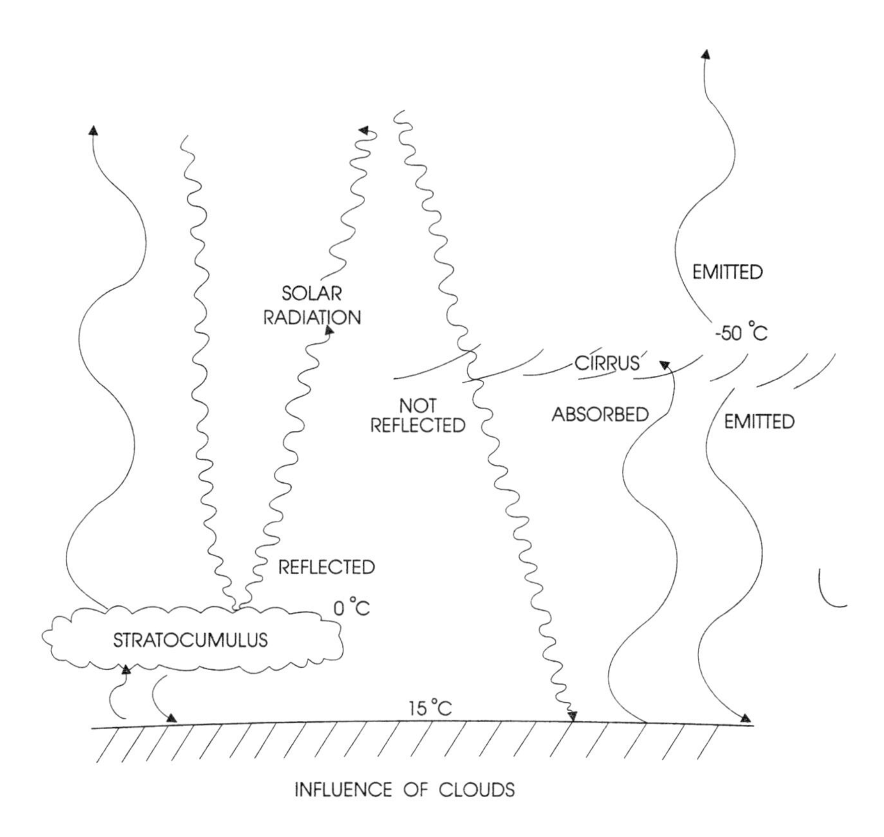


Fig. 14. The influence of clouds on the climate is quite complex. Generally, thick clouds, such as stratocumulus, reflect much more solar radiation than do thin clouds, such as high cirrus clouds. Reflecting solar radiation tends to cool the climate. All clouds absorb Earth's infrared radiation and effectively prevent infrared radiation from below being lost to space (they are acting as very strong greenhouse gases). The overall impact as a greenhouse depends on the temperature of the upper surface of the clouds, the surface that is radiating to space. If the top of the cloud is very high, its temperature will be very cold and it will radiate a small amount of energy to space. A low cloud top will be warmer and radiate more energy to space. Hence, high clouds are more effective in blanketing the earth than low clouds. In a changed climate, the redistribution of clouds may make clouds more or less effective as reflectors of solar radiation or as a greenhouse blanket.

although models of finer grid are now being used in some countries. Major topographic features are smoothed and some do not even appear. Climate models are limited by computer capacity to deal only with the larger-scale aspects of climate. Once some confidence has been gained that the model can simulate the major features of present climate, then its sensitivity to a change in greenhouse gas concentrations can be determined. Two approaches have been used. Firstly, model results for carbon dioxide of 300 ppmv are compared with the results of 600 ppmv; the differences are due to doubling carbon dioxide, since all other parameters in the model are kept constant. These are called doubled-$CO_2$ experiments and are climate scenarios as described earlier. The impacts of the other greenhouse gases are included through their equivalent carbon dioxide radiation forcing. Doubled-$CO_2$ experiments are tests of the sensitivity of the model to doubled carbon dioxide. If we have some confidence that the model simulates the present climate and has characteristics consistent with our understanding of how the climate system functions, then the sensitivity of the climate model should approximate the sensitivity of the real climate system.

The 1990 IPCC Scientific Assessment summarized the results of many global climate models for the sensitivity of climate to a doubled atmospheric concentration of greenhouse gases (some are reproduced in Table 4). They then concluded that the equilibrium global mean temperature would change between 1.5 and 4.5°C, with a 'best estimate' of 2.5°C, for doubled atmospheric carbon dioxide concentration.

The second approach, which is now being used in several centres, is to gradually increase the atmospheric carbon dioxide concentration, at some rate consistent with emission scenarios; these are called *transient simulations*. They are clearly more realistic but require a realistic ocean model and much-improved computational power and time.

Although the IPCC consensus was based on the input and review by several hundred scientists, some scientists do not support its conclusions. These scientists claim that the climate is much less sensitive to changing greenhouse gas concentrations and that no significant climate warming should be expected over the next century, regardless of emissions to the atmosphere. The main argument is that climate models do not properly represent the complex processes of the global water cycle and clouds. Another part of the argument rests on the fact that the global warming over the past century (about 0.5°C) could be entirely due to natural variability. They argue that, if the climate is as sensitive as suggested by IPCC, there should already be more warming since carbon dioxide concentrations have increased by more than 25%.

Table 4. Examples of modelled changes in global mean temperature ($T$) and precipitation ($P$) corresponding to doubled atmospheric concentration of $CO_2$. Each model consists of a comprehensive atmospheric model coupled to a mixed-layer ocean model.

| Model | $\Delta T$ (°C) | $\Delta P$(%) | Comments |
|---|---|---|---|
| GFDL | 4.0 | 8 | Geophysical Fluid Dynamics Laboratory, Princeton |
| | 4.0 | 8 | Similar to above, but higher resolution |
| UKMO | 5.2 | 15 | UK Meteorological Office |
| | 2.7 | 6 | Changed cloud water scheme |
| | 1.9 | 3 | Variable cloud radiative properties |
| | 3.5 | 9 | Higher resolution |
| CCC | 3.5 | 4 | Canadian Climate Centre, high resolution |

There are strong counter arguments. One has already been mentioned; the destruction of ozone by CFCs has been reducing the greenhouse effect by reducing the ozone contribution.

The second concerns atmospheric aerosols of anthropogenic origin. The aerosols in question result from emissions of sulphur dioxide to the atmosphere which form small sulphate aerosols. Sulphate aerosols have two impacts on radiation transfer in the atmosphere. They scatter incoming solar radiation and they change the properties of clouds so that clouds reflect more solar radiation. Both processes cool the climate. Although these general influences have been recognized for some time, recent observations and radiation computations show that the present distribution of aerosols decreases the incoming solar radiation by an amount that approximately balances the computed increase in terrestrial radiation due to the increased greenhouse effect, when averaged over the past decade and over the globe. These results, as with those on the CFC–ozone radiative reductions, have been further clarified in the last few years and are now included in the transient simulations of climate change (see Figs. 11, 13 on pp. 47, 49, and Kuemmel, 1996). An additional point is that the large inertia of the oceans will slow down the rate of climate change: transient models are only beginning to simulate this effect.

Because sulphate aerosols are contributors to acid rain, international action is being taken to reduce their emissions into the atmosphere. Since sulphate aerosols have a relatively short lifetime in the atmosphere, they do not accumulate in the atmosphere and their concentration will decrease rapidly in the future as emission controls become effective. Similarly, since CFCs destroy the ozone layer which shields us from harmful ultraviolet radiation, actions are being taken to reduce their emissions. Hence, although these processes probably partially counterbalance the increased greenhouse effect at present, this will not be the case in the future.

The IPCC Scientific Assessment also concluded that the high northern latitudes will warm by about twice the global mean, tropical latitudes will warm less than the global mean and land areas will warm more than the oceanic areas. Changes at high southern latitudes are less clear but are likely to be small. However, a caution was added: surprises in the climate system are possible and unknown biospheric and other feedbacks could cause more rapid and/or larger climate change. The timing of climate change corresponding to the changing atmospheric concentrations of greenhouse gases will depend on the slowest component of the climate system, i.e., the oceans. The oceans will cause a lag and/or reduction in climate change since they absorb heat in two ways. One is absorption only in the upper layers of the ocean with a slow, rather inefficient diffusion of heat to depth. Initially, the rate of climate change would be very small; however, after a lag of time ($L$) (Figure 15), the rate of change of the whole system will be the same as the fast component. If the ocean acts in this way, we should see rapid climate change in the next century, after a lag of a few-to-several decades. A second way for the ocean to absorb heat is through convection at high latitudes which acts to warm a much deeper layer of the ocean and effectively reduce the rate of climate change by some fraction ($R$).

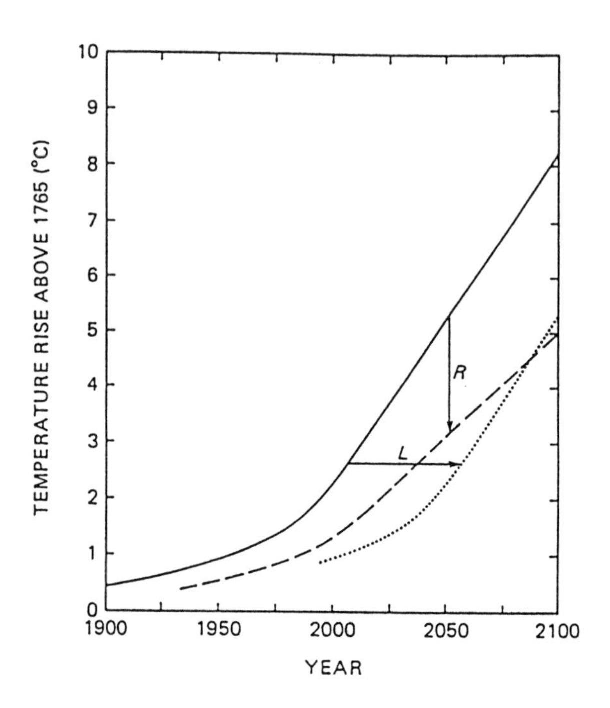


Fig. 15. Schematic of the role of the oceans in modifying global climate change. Assume the response of the fast climate system is as shown by the solid line. If the ocean functions just to delay climate change, the response of the coupled climate system will be as shown by the short-dashed line, with a delay or lag (*L*). If the ocean functions to reduce climate change by some factor (*R*), then the coupled system response will be as shown by the long-dashed line.

Simulations with atmospheric climate models coupled to a multi-layer, coarse-resolution ocean model have been used with the atmospheric concentration of carbon dioxide increasing at rates similar to 1% per year. After an initial period, global mean temperature increased approximately linearly. The temperature changes varied regionally (see Figure 16), but the patterns were generally similar to the fast climate model results discussed above, except for being much smaller in the North Atlantic around 60°N and for the Antarctic Circumpolar Current. In both these areas there is deep convection and the result is much less local warming. As the climate warms in coupled ocean-atmosphere models, increasing precipitation over the North Atlantic freshens the surface waters, reduces their density, and decreases deep convection. Eventually, deep convection stops, leading to accelerated climate warming. Since deep convection is also responsible for removal of carbon dioxide from the surface waters, reduction of deep convection would reduce the rate of uptake of carbon dioxide by the ocean and allow the rate of increase in the atmosphere (for the same rate of emissions) to accelerate. Through these processes, the ocean may provide a strong positive feedback on greenhouse gas warming of the climate. There is evidence that changes of this type may have occurred naturally in the past.

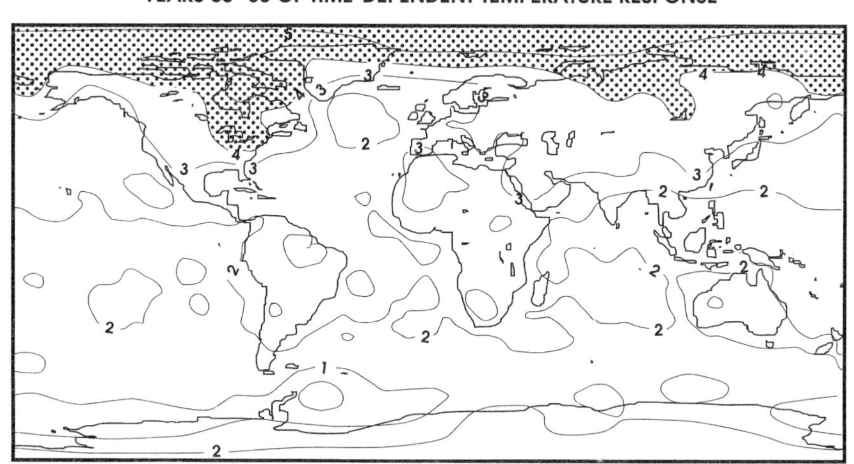


Fig. 16. The regional distribution of warming, based on the annual average temperature change for the period corresponding to doubled atmospheric carbon dioxide concentration. The results are from the time-dependent stimulation with atmospheric $CO_2$ increasing at 1% per year (compounded) of the US Geophysical Fluid Dynamics Laboratory coupled ocean–atmosphere climate model.

Based on the emission scenarios and the use of the best climate models for the sensitivity and response of the climate system, the IPCC predicted (Figure 17; Table 3): Scenario A leads to a increase in global mean temperature of about 1°C above present values by 2025 and 3°C before 2100. For each scenario, high northern latitudes would warm by several times more during the winter and generally land areas will warm more than the oceans. The global mean sea level is predicted to rise about 6 cm per decade over the next century (with an uncertainty of 3–10 cm per decade). This corresponds to a rise of about 20 cm by 2030 and 65 cm by 2100. Changes for other scenarios would be less. The temperature and sea level rises would not be steady because other, natural factors will modulate the change. Presumably, the timing of changes in precipitation will be similar, but we have less confidence in the results. For the A Scenario the temperature change will be greater and faster than any seen over the past 10 000 years.

The 1995 Scientific Assessment of the IPCC has used different (but overlapping) scenarios of emissions), as noted above, and included the role of aerosols. For a mid-range emissions scenario and the best estimate of climate sensitivity, models project a global temperature increase, relative to the present, of 2.0°C by 2100. In all cases, the warming will continue after 2100 due to the inertia of the oceans.

Global climate models also show that the precipitation will change (between 3% and 15% in terms of global mean for doubled-$CO_2$; Table 4, p. 86) but models differ widely in their prediction of regional variations. In many respects, the most important weather element in terms of its impact on ecosystems and human activities is precipitation. Excess rain and snow can create floods and avalanches while droughts can be just as serious. An

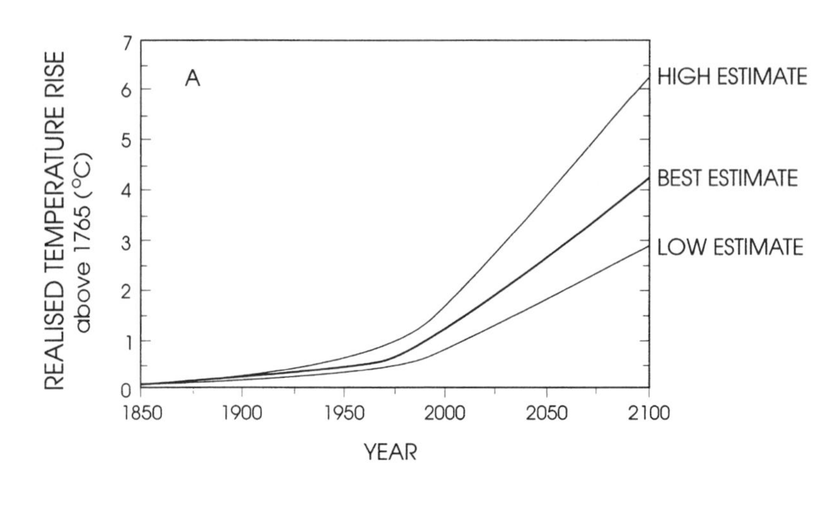


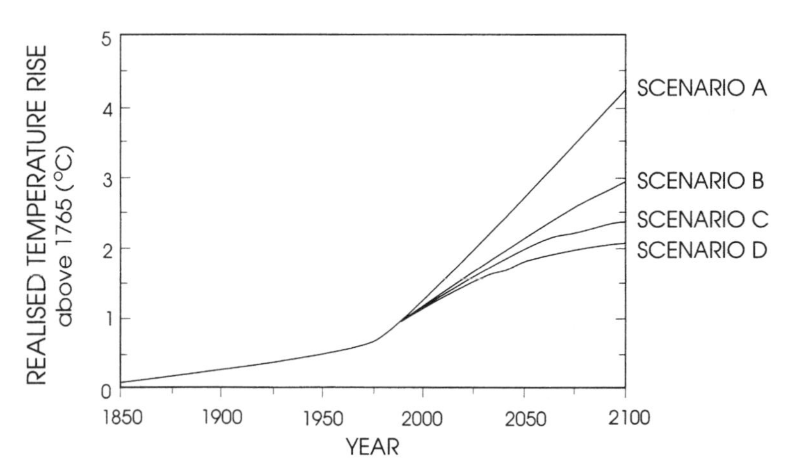


Fig. 17. The IPCC global warming scenarios. The realized temperature rise (above the year 1765, before the industrial revolution) for the A emissions scenario (upper) with high and low estimates and for emissions scenarios A, B, C and D (below).

important parameter of practical interest is the difference between precipitation and evaporation, which we might call available water. In many places, the available water is a small difference between two much larger numbers. Suppose we have a climate where the mean annual precipitation is 500 mm and the mean annual evaporation is 450 mm; the available water for runoff and replenishing soil moisture will be 50 mm. With climate change, the precipitation may increase by, for example, 5% (to 525 mm) and the evaporation by 10% (to 495 mm). The result is a decrease in available water to 30 mm, a change of 40%. Changes to available water, in regions where there are already water supply problems, may be catastrophic.

Large-scale changes in soil moisture accompanying climate change are only approximately treated in climate models. However, it was found that, for doubled $CO_2$, soil moisture in the Great Plains of North America for July and August was about half of values for the simulation of present climate. Although winter rainfall increases, spring and summer rainfalls decrease due to changes in atmospheric circulation. Snowmelt, which is a major source of spring moisture, occurred earlier with the warmer climate and resulted in more spring runoff. Evaporation increases through the winter and spring due to higher temperatures and increased solar radiation absorption. Changes in snow storage/runoff relationships will be very important for many regions.

We again must stress the possibility of surprises. The biosphere has been neglected in these calculations and could amplify or possibly reduce the change. The direct effect of aerosols could be to reduce the change while indirect impacts through clouds are less clear. Actual greenhouse gas-induced warming will be delayed by thermal inertia of the oceans, but will also continue long after the composition of the atmosphere is stabilized.

Another kind of surprise is related to possible ‘flip-flops’ in the deep ocean conveyor belts. Recent results from two ice cores drilled in central Greenland reveal large abrupt climate changes towards the end of the last ice age, of at least regional extent, believed to be associated with sudden changes in the currents of the North Atlantic Ocean.

## Potential Impacts of Global Warming

The IPCC has summarized the potential impacts of climate change. It concluded that there would be impacts on agriculture (especially in arid and semi-arid areas), forests, natural terrestrial ecosystems, hydrology and water resources, human settlements, energy, transportation, industry, human health and air quality, and the oceans, coastal zones and fisheries; in other words, in most natural resource-based economic sectors. Climate change, at the rate and magnitude currently estimated over the next 40–50 years, may exceed the ‘critical loads’ for certain ecosystems and economic sectors, and may widen the gap between developing and developed countries, because the impacts in many cases will be most severely felt in regions already stressed.

For agriculture, there are potential benefits but in many areas, some already under stress, a warmer world will aggravate problems of water shortage, pests and poor soil. Natural forests and ecosystems are not as adaptable and climatic zones may become unmatched to species that have developed over thousands of years. While forest management practices may be amenable to overcoming such difficulties, fully natural ecosystems may not have time to adjust naturally to change, and may perish. Water availability may be the most critical question. Small changes in precipitation and evaporation can lead to large changes in available water. Human settlements and agriculture have developed in areas with very precarious water supplies; climate warming will generally intensify these difficulties.

A warmer climate is expected to change the patterns of disease and pests, shifting them to higher latitudes and putting large populations at risk. Climate change could also initiate large human population migrations, leading to social instability in some areas. Mean sea level will rise due to changes in water volume (through additions from land glaciers and by thermal expansion) or coastal subsidence. This magnitude of sea level rise would seriously threaten low-lying islands and coastal zones. Coastal areas are inherently hazardous areas, which at the same time, continue to increase their human populations and economic activities. A 1 m rise would result in, for example, 17% of Bangladesh's arable land being inundated and large displacements of people in Indonesia and Vietnam. Flat deltaic areas are most vulnerable. Coral atoll nations are also at risk. In addition, sea level rise would affect fresh water supplies and agricultural productivity of these countries through both inundation and salt water intrusion. Rising sea level increases the risk of flooding, and there is an increased likelihood of river surges. Damage to coastal infrastructures could be in the hundreds of billions of dollars. There will be a general physical deterioration of coastal regions due to erosion, inundation and recession of barrier islands, coastal atolls and other shorelines. Many recreational beaches would be eliminated and costs of maintaining navigation channels would be greatly increased. An increase in oceanic temperatures would result in a northward shift of ocean-climate zones with impacts on fish and other species.

Added to the stresses to society and the biosphere that global environmental change is likely to cause, there is the compounding effect of increasing populations and economic activity, particularly in the coastal zones of many developing countries. Even during the last 30 years, there has been an increase in economic losses and insured economic losses from natural disasters, as shown in Figure 18. This trend is of considerable concern to the insurance business.

## Climate Change Research in Service to Society

Climate change research provides a service to society, particularly with respect to its desire to achieve long-term sustainability. To effectively deliver this service, we need to determine the rate and magnitude of climate warming, the impact on the global hydrological cycle and the change in the frequency and severity of extreme events, including their regional variations. The World Climate Research Programme is aimed at providing a quantitative understanding of climate and prediction of climate change on all times scales. The support of all governments is essential in order to understand and eventually predict the evolution of the Earth's climate.

Environmental issues, particularly climate variability and change, are of concern to society and climate research provides the basis for developing of policy. The impacts of climate variability and change on resources and our economic structures add to the difficulty of attaining sustainable development. Public and industry acceptance of either

**WILL CLIMATE CHANGE BRING SIGNIFICANT IMPACTS?**
(Some of the views of IPCC Working Group 2,1995)

* Human health may be seriously affected, with the spread of tropical diseases into the temperate zones, and with urban heat-related deaths increasing several-fold.
* Natural ecosystems will be greatly stressed, particularly coral reefs and the boreal forest.
* Increased fire frequencies and pest outbreaks are likely to decrease average age, biomass and carbon store of forests.
* On aggregate, agricultural production is not likely to be greatly affected, with slight gains in countries such as Canada but with locally very significant losses in many developing countries.
* Coastal zones and small islands are particularly vulnerable to sea level rises. A 50 cm rise in sea level would increase the number of people world-wide subject to coastal inundation from 46 to 92 million.

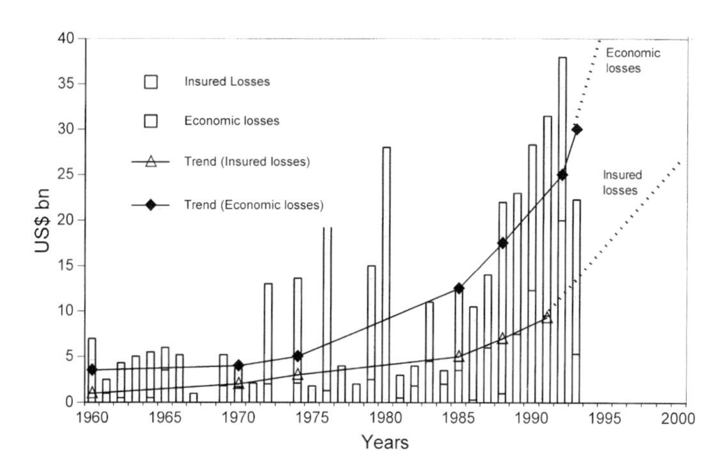


Fig. 18. Global losses from natural disasters 1960–1993. From F. Pisano (1995) Natural disaster reduction and insurance markets, IDNDR *STOP Disasters*, **24**, 21–22.

increased costs or regulation to mitigate effects of future climate change is far from certain. While the public and industry have already demonstrated a renewed and widely-based concern about such environmental impacts, it would be unreasonable to expect their support for changes in current economic patterns without clear, well-founded, and widely disseminated scientific evidence of the negative consequences of no action.

Many industries (particularly in the forest, fisheries and agriculture sectors) will be directly impacted by climate variability and change, through effects on the resources they harvest. Other companies will be directly affected by national commitments to alter our energy system (such as petroleum, electric power, and mining companies). Climate prediction offers the possibility of improved strategic planning by governments and industry leading to better efficiency, reduction of waste and less impact of climate variability and change on society.

Appendix 3 includes a discussion of the methods that are used for assessing the effects of climate change on the biosphere and society. Appendix 5 is an inventory of adaptation strategies relating to climate change.

## Summary

The climate system is a complex, interacting integration of the atmosphere, oceans, land surfaces and their vegetation, and snow and ice, which varies on a wide range of space and time scales – ranging from months to millennia. Now, for the first time, human activities are starting to influence climate on the global scale. Climate variability and change have important impacts on society and some societies and functions are much more vulnerable than others. Information on climate variability and change (i.e. predictions) has value and provides a basis for development of response strategies, including abatement of emissions, adaptation and others. Internationally-coordinated national research efforts are the foundation for developing understanding and prediction capability.

For most purposes, global mean temperature change is not a very useful variable; regional climate predictions are needed. We also need to predict the changes in the water cycle: precipitation, evaporation, available water, river flow, etc. And, we need to know the rise in sea level. Each of these scientific issues is being addressed in the World Climate Research Programme or other programmes.

Based on scenarios of future emissions of carbon dioxide and other greenhouse gases to the atmosphere, climate models are predicting significant global climate warming. For the mid-range emissions scenario, including aerosols and using the best estimate of climate sensitivity, the global climate is predicted to warm by 0.2°C per decade or about 2°C by the year 2100. Such a temperature change would be unprecedented over the last 10 000 years and would have major impacts on ecosystems, sea level and climate-sensitive activities. There is uncertainty in this prediction due to our imperfect understanding and ability to model the climate system. However, when the costs of taking action on reducing emissions of greenhouse gases are compared with the risks in taking no action, an emissions reduction strategy makes sense. The question of risk management must be addressed through comprehensive analysis of climatic, economic and social factors in an integrated way.

## Selected References

ICSU (1992) *An Agenda of Science for Environment and Development into the 21st Century*. Dooge, J. C. I., Goodman, G. T., la Rivière, J. W. M., Marton-Lefevre, J., O'Riordan, T., and Praderie, F. (eds). Cambridge University Press, Cambridge. 331 pp.

IPCC I (1990) *Climate Change; The IPCC Scientific Assessment.* Houghton, J. T., Jenkins, G. J., and Ephraums, J. J. (eds). Cambridge University Press, Cambridge. 364 pp.

IPCC I (1992) *Climate Change 1992: The supplementary report to the IPCC Scientific Assessment.* Houghton, J. T., Callander, B. A., Varney, S. K. (eds), Cambridge University Press, Cambridge. 200 pp.

IPCC I (1994) *Climate Change 1994: Radiative Forcing of Climate and an Evaluation of the IPCC IS 1992: Emission Scenarios.* Houghton, J. T., Meira Filho, L. G., Bruce, J. P., Hoesung Lee, Callander, B. A., Haites, E. F., Harris, N., and Maskell, K. (eds). Cambridge University Press, Cambridge.

IPCC I (1996) *Climate Change 1995: The Science of Climate Change.* Houghton, J. J., Meira Filho, L. G., Callander, B. A., Harris, N., Kattenberg, A., and Maskell, K. (eds). Cambridge University Press, Cambridge, 584 pp.

IPCC II (1990) *Climate Change; The IPCC Impacts Assessment.* Tegart, W. J. M., Sheldon, G. W. and Griffiths, D. C. (eds). Australian Government Publishing Service, Canberra, 245 pp.

IPCC III (1991) *Climate Change; The IPCC Response Strategies.* Bernthal, F. M. (Chair). Island Press, Washington, 272 pp.

IPCC II (1996) *Climate Change 1995: Impacts, Adaptations, and Mitigation of Climate Change – Scientific–Technical Analyses.* Watson, R. T., Zinyowera, M. C. and Moss, R. H., (eds). Cambridge University Press, Cambridge, 880 pp.

IPCC III (1996) *Climate Change 1995: Economic and Social Dimensions of Climate Change.* Bruce, J., Lee, H., and Haites, E. (eds). Cambridge University Press, Cambridge, 464 pp.

Jaeger, J. and Ferguson, H. L. (eds) (1991) *Climate Change: Science, Impacts and Policy.* Cambridge University Press, Cambridge, 591 pp.

Kuemmel, B. (1996) Editorial essay, *Climatic Change,* **32**, 379–385.

# 5. Policy Responses to Global Environmental Issues: An Introductory Overview

J. C. I. DOOGE
*President of ICSU*

## Chronology of Events Mentioned in this Chapter

1979: First World Climate Conference
1980: Launching of the World Climate Programme by WMO, UNEP and ICSU
1983: Establishment of the UN Brundtland Commission by the UN
1985: The Villach Conference on the Assessment of the Role of Carbon Dioxide and Other Greenhouse Gases in Climate Variation and Climate Change
1987: Release of the Brundtland Commission Report and adoption by the UN
1990: Second World Climate Conference and Ministerial Sessions
1991: ICSU Conference ASCEND 21 (An Agenda of Science for Environment and Development into the 21st Century)
1991: Release of the Business Charter for Sustainable Development by the International Chamber of Commerce
1991: Rotterdam Declaration of the Second World Industry Conference on Environmental Management (WICEM-II)
1992: The UN Rio Conference UNCED (United Nations Conference on Environment and Development)
1992: Publication of the book *Changing Course* (S. Schmidheiny, ed.) by the Business Council for Sustainable Development
1993: Declaration of the Business Council for Sustainable Development

## New Approaches: New Partnerships

The preceding chapters have outlined the extent of our scientific knowledge about the current environmental state of planet Earth and the likely prognosis for the next century. Subsequent chapters will deal with the responses to that situation by international inter-

*R. E. Munn, J. W. M. la Rivière and N. van Lookeren Campagne (eds), Policy Making in an Era of Global Environmental Change, 97–112.* 

**How did 'biogeochemical cycling' become a global policy issue?**

How did this happen? A press release (October 1979) made jointly by Mostafa Tolba, Executive Director of UNEP, and Gilbert White, President of SCOPE, 'drew attention to the fundamental scientific importance of understanding the biogeochemical cycles which link and unify the major chemical and biological processes of the Earth's surface and atmosphere' and invited 'members of the scientific community in the various disciplines to contribute to the design and execution of a collective endeavour'. The press release points out that several important global environmental issues (climate change; stratospheric ozone depletion; acidic deposition; agricultural land degradation) are inter-related through the cycling of carbon, nitrogen, phosphorus and sulphur. 'Understanding the working of biogeochemical cycles is made much more difficult – and more necessary for policy purposes – by the evidence that each of the cycles influences the others, and may be related to trace elements and toxins'.

governmental organizations, national governments, NGOs and business interests. The present chapter provides an introductory overview of the interactions that have occurred up to now and the types of dialogue and partnerships that will be required in the future.

The past decade has seen a remarkable transformation of the relationships between the scientific research community working on global environmental processes and people involved in policy formulation, in both the public and the private sectors. These changes are illustrated by the contrast between the First World Climate Conference (February 1979) and the Second World Climate Conference (November 1990). At the 1979 Conference, the vast majority of the participants were scientists, and discussions were devoted to consideration by natural scientists of the nature of climate change, and by natural and social scientists of the likely impacts of such change. The Second World Climate Conference, on the other hand, was much broader in scope and was divided into two parts: (1) scientific and technical sessions; and (2) ministerial sessions. In the scientific and technical sessions, there was greater emphasis on socio-economic impacts than at the 1979 Conference, and one of the four panel discussions was concerned with the responses of industry to global warming (see Box 1).

In the period between the two World Climate Conferences, scientists and decision-makers became increasingly concerned about the problem of possible global warming. The World Meteorological Organization (WMO), the United Nations Environment Programme (UNEP) and the International Council of Scientific Unions (ICSU) co-operated in the planning and implementation of the World Climate Programme and in the organization of the 1985 Villach Conference on the Assessment of the Role of Carbon Dioxide and Other Greenhouse Gases in Climate Variation and Climate Change.

The 1985 Villach Assessment was particularly timely, feeding into the deliberations of the World Commission on Environment and Development (the so-called Brundtland Commission) established by the United Nations in 1983. The 1987 Report of the Brundtland Commission gave rise to a new awareness by individual governments and

**Box 1. From the Proceedings of the Second World Climate Conference (arising from the deliberations of a Panel concerned with the response of industry to global warming)**

Protection of the environment is a key issue facing industry in the 1990s. But the particular issue of global warming, or more precisely the climatic changes associated with global warming, is one of the least certain and most complex threats facing our planet in both timing and effect.

Industry recognizes the degree of concern which now exists on this issue based on current knowledge. It concludes that there are a number of policy responses and actions which can be justified in their own right, can be initiated relatively quickly, and would make a significant reduction of $CO_2$ emissions. These would provide momentum while the scientific analysis continues and the ground can be prepared for more significant changes in the policy if these are eventually judged to be necessary.

While industry accepts the challenge to help protect and improve our planet, it is but one of the participants, and can make its contribution only with the help and consent of other parties. Insurance against global warming will have to be collective, not individual. If the world decides to accept significant costs now to insure itself against potentially higher future costs, it will be primarily through a political process.

Industry's approach should be based on three principles: good science, sound economics and a proper dialogue.

Page 493, *Climate Change: Science, Impacts and Policy* (Jäger and Ferguson, 1991)

intergovernmental bodies of global problems such as climate change, and many follow-up actions were subsequently taken. For example, WMO and UNEP in 1988 jointly set up the Intergovernmental Panel on Climate Change (IPCC) to assess more closely the current state of knowledge on climate change.

**From the 1985 Villach Conference Statement**

As a result of the increasing concentrations of greenhouse gases, it is now believed that in the first half of the next century, a rise of global mean temperature could occur which is greater than any in man's history.

See WMO publication no. 661.

Similar developments were taking place at about the same time in other sectors of society. In business and industry, for example, there was a growth of awareness of global issues and it was realized that the linear model of industrial development (research followed by development, followed by manufacture, followed by sales) was no longer appropriate and must be replaced by an interactive model covering the entire life cycle of the product and allowing for appropriate feedbacks amongst all the components involved.

The new situation in relation to global environmental problems called for new approaches by all groups involved – research scientists, governments and the business community. This issue was brought into focus by the need for all three groups to make appropriate inputs into the UN Conference on Environment and Development (UNCED) held in Rio de Janeiro in June 1992. ICSU, as principal scientific advisor to the UNCED preparatory process, organized an International Conference (November 1991) to develop an Agenda of Science for Environment and Development into the 21st Century (ASCEND 21). The conference proceedings define the basic problems, summarize our understanding of the Earth System and discuss strategy and response options. The Conference also 'forcefully asserted the responsibility of science (encompassing the natural, social, engineering and health sciences) to provide independent explanations of its findings to individuals, organizations and governments'.

At the same time, a group of some 50 business leaders under the chairmanship of Stephan Schmidheiny created the World Business Council for Sustainable Development and acted as principal advisor for business and industry in the preparations for UNCED. In its 1993 Declaration, the Council recognized the responsibilities of industry and emphasized the need for 'new forms of cooperation between government, business and society'.

Endorsement of the Rio Declaration and a text for Agenda 21 by 179 countries, and the acceptance of new mandates by the scientific and business communities have set the stage for creating new institutional mechanisms to meet the global challenges. Much needs to be done, of course. Some of the important considerations to be taken into account in finding a way forward are discussed in the remainder of this chapter.

## Sustainable Development

### *The Brundtland Report*

The rise in both world population and the consumption per head has led to pressure on the global environment of an intensity unknown in human history. This calls for re-examination of assumptions about the best means of optimizing growth and welfare, assumptions that were taken as obviously true at an earlier stage of human development.

In the last 50 years there has been a transformation from a world relatively empty of human beings and man-made capital to a world relatively full of people and capital. Daly (see Box 2) attributes the failure of decision-makers and most economists to recognize this shift to the difficulty in detecting the early stages of exponential growth and to an over-emphasis in classical economics on the substitutibility of limiting production factors.

A key change in thinking over recent years has been the realization that it is possible to harmonize developmental and environmental considerations rather than to treat them as inherently antagonistic. This view was elaborated by the Brundtland Commission,*

* The official title of the Commission was the United Nations World Commission on Environment and Development.

**Box 2**

The evolution of the human economy has passed from an era in which man-made capital was the limiting factor in economic development to an era in which *remaining* natural capital has become the limiting factor. Economic logic tells us that we should maximize the productivity of natural capital and its total amount, rather than increase the productivity of man-made capital and its accumulation, as was appropriate in the past when it was the limiting factor.

Herman Daly (1992) From empty-world economics to full-world economics, in *Population, Technology and Society* (Island Press, Washington), pp. 23–37.

which gave the phrase 'sustainable development' its current high visibility. The 1983 terms of reference for the Commission were to:

(1) Define shared perceptions of long-term environmental and development challenges, and the most effective methods to respond to them.
(2) Recommend means to foster greater cooperation among developed and developing countries, and to maintain mutually supportive objectives, taking into account the interrelationships among people, resources, environment and development.
(3) Propose long-term strategies to achieve sustainable development, combining global economic and social progress with respect for natural systems and environmental quality.

In its final report (the monograph *Our Common Future*), the Commission rejected the assumption that the goals of environmental protection and economic development are incompatible, and argued that neither environmental protection nor economic development is sustainable without proper attention to the other. The Commission defined *sustainable development* as development that meets the needs of the present without compromising the ability of future generations to meet their own needs, and recommended it as the basis for prudent management into the twenty-first century and beyond. The Commission Report recognized that sustainable development requires changes in the domestic and international policies of every country. It is for each nation to work out its own policies but it is clear that the transition to sustainable development requires joint management by all peoples.

### *Responses of Government Bodies to the Brundtland Report*

Implementation of programmes that lead society towards sustainable development is difficult for several reasons, the two most important being:

(1) The phrase *sustainable development* is an aspiration, which is hard to interpret in practical terms.

(2) The idea of intergenerational equity involves value judgements, particularly with respect to discounting the future, and with respect to changes over decades in codes of ethics and societal 'wish lists'.

> When the environmental costs of economic activity are borne by the poor, by future generations, or by other countries, the incentives to correct the problem are likely to be weak (Arrow *et al.*, 1995).

Nevertheless, the reaction to the Brundtland Report by governments, the business community and the scientific community has been positive and constructive. In 1987 the General Assembly of the United Nations passed a resolution adopting the Commission's Report as a guide for UN operations and commending it to governments. In 1989 the G7 summit called for 'the early adoption worldwide of policies based on sustainable development'. Subsequently, the 1992 Rio Declaration on Environment and Development (adopted by 179 countries in June 1992), contains the following:

> The right to development must be fulfilled so as to equitably meet developmental and environmental needs of present and future generations (Principle 3);
>
> In order to achieve sustainable development, environmental protection shall constitute part of the development process and cannot be considered in isolation from it (Principle 4).

Chapter 8 of the Rio document, Agenda 21, discusses the implementation of these principles (pp. 65–74).

*Responses of the Business Community to the Brundtland Report*

The principle of sustainable development has been strongly endorsed by representatives of the business community. A Business Charter for Sustainable Development was prepared by a task force of the International Chamber of Commerce. This Charter was launched at the Second World Industry Conference on Environmental Management in 1991, and was subsequently endorsed by hundreds of individual enterprises in all parts of the world. At the 1991 Conference the participants accepted the commitment to sustainable development in these words:

> We support the principles of sustainable development which go beyond the present concept of environmental excellence. It involves a process of integrating environmental criteria into economic practice, satisfying basic needs, and conserving nature's capital for future generations.

In 1992 the International Chamber of Commerce published a 351-page report amplifying the context and implications of the Business Charter for Sustainable

Development. There has also been action by business organizations in individual countries. Thus, for example, Keidanren, the influential business group in Japan, adopted a Global Environmental Charter in 1991. International associations for particular industries have drafted new codes of conduct to take account of the new situation.

Some 50 business leaders were convened specifically to provide a business input to the Rio Conference. The Declaration of this Business Council for Sustainable Development contains the following words:

> Business will play a vital role in the future health of this planet. As business leaders, we are committed to sustainable development, to meeting the needs of the present, without compromising the welfare of future generations (first paragraph).
>
> New forms of cooperation between government, business and society are required to achieve this goal (third paragraph).

Following the Rio Conference, the International Chamber of Commerce and UNEP formed a joint Advisory Panel, whose goal is to encourage business enterprises worldwide to implement the Business Charter for Sustainable Development.

It is clear that business leaders have reacted in a responsible and supportive way.

*Responses of the Scientific Community to the Brundtland Report*

Reference has already been made to the role of ICSU as principal scientific advisor to the process leading up to Rio. Through its President, Professor M. G. K. Menon, ICSU pledged its support for a number of follow-up activities, including the decision:

> To strengthen further its capacity to play its role in the evolving partnership among science, governments, international organizations and business and industry.

The response of the engineering profession was also prompt and positive. The Fédération Internationale des Ingineurs Conseils (FIDIC) established an Environmental Task Committee following publication of the Brundtland Report, which resulted in the adoption of a Policy Statement (June 1990) on Consulting Engineers and the Environment. The Statement stressed that:

> Engineers should provide leadership in achieving sustainable development. Consulting engineers should combine their traditional skills with broader applications of physics, chemistry, biology and other disciplines to lead interdisciplinary teams directed at achieving environmental solutions.

The Policy Statement was subsequently adopted by a number of national member organizations of the Federation.

## Economic Considerations

A key source of discussion between industry and environmentalists is the relation between economy and environment with respect to sustainable development. In the 1992 volume organized by the Business Council for Sustainable Development and edited by Schmidheiny are found the assertions that:

> The cornerstone of sustainable development is a system of open, competitive markets in which prices are made to reflect the cost of environmental as well as other resources (opening paragraph of Chapter 2).
>
> Market economies must now rise to the challenge and prove that they can adequately reflect environmental truth and incorporate the goals of sustainable development (last paragraph of Chapter 2).

Realization of the latter goal is far from easy. The members of OECD agreed over 20 years ago to the 'polluter pays principle', but full implementation has proved difficult.

The need to include the real cost of natural resources gave rise in the 1970s to the development of environmental economics, which concerns itself with means of broadening the scope of neo-classical economics to include environmental factors. The roots of such an approach can be traced back in time for well over a century. Thus John Stuart Mill in his Principles of Political Economy, first published in 1848, spoke of the need for dealing with so called 'free goods':

> The earth itself, its forests and waters above and below the surface ... are the inheritance of the human race. What rights, and under what conditions, a person shall be allowed to exercise over every portion of this common inheritance cannot be left undecided. No function of government is less optional than the regulation of these things, or more completely involved in the idea of a civilized society.

As is well known, the free market mechanism with equilibrium prices has certain optimal properties. But there are many assumptions that have to be fulfilled in order to ensure these properties. A fundamental assumption is that there be no collective (or external) side effects of production or consumption, in addition to what individuals consider to be of importance to them.

If collective side effects (externalities) are substantial and important, the classical doctrine of the blessings of free trade simply becomes irrelevant as a guideline for economic policy. This is a conclusion that any serious student of economics can verify by means of standard economic textbook theory.

Haavelmo, T. and Hansen, S. (1992) On the strategy of trying to reduce economic inequality by expanding the scale of human activity, Chapter 3 of *Population, Technology and Society*, edited by R. E. Goodland, H. E. Daly and S. El Serafy (Island Press, Washington) 156 pp.

Three of the major tools used in environmental economics are:

(1) *Benefit–cost analysis*, which has been used in the evaluation of certain types of large-scale civil engineering works for over 50 years. Despite this long experience, benefit–cost analysis still retains a degree of subjectivity due to: (a) the presence of indirect and extra-market elements in both benefits and costs; (b) the difficulty of forecasting certain factors over the time-scales involved; and (c) the critical effect of the choice of a discount rate.

(2) *Cost-effectiveness analysis*, which seeks to minimize cost of meeting a prescribed environmental objective. This method also has innate elements of subjectivity both in relation to the setting of environmental objectives and the minimum cost analysis itself.

(3) *Resource accounting*, which attempts to compute national sets of resource accounts, providing a basis for comparisons with national economic accounts.

The impetus for introducing environmental economics into traditional economics came largely from economists working in the renewable resource field. In the last decade environmental economics has been superseded to some extent by 'ecological economics', a response to the Brundtland challenge, and to long-term concerns about the carrying capacity of the Earth. The current status of *ecological economics* is reflected in the journal *Ecological Economics* and in the activities of the International Society for Ecological Economics. See also Costanza (1991) and Arrow *et al.* (1995). In essence, ecological economics is about Herman Daly's full-world economics in contrast to earlier 'empty-world' economics, and the new approaches go far beyond conventional economic calculations of environmental damage. In the 'classical' approach, the cost of reducing air pollution emissions is compared with the costs of more frequent dry cleaning, more frequent painting of houses, replacement of rusty metals, hospitalization of asthmatics, etc. In the new ecological economics, short-term benefits and long-term costs are compared, e.g., the current market value of a forest vs. its long-term economic, social and ecological benefit. Daly and Cobb (1989) have used the metaphor that even the sturdiest ship will eventually sink if the load is too big. There is little comfort in the fact that the load was optimally allocated and fairly distributed at the time of the sinking.

> The market works well but not all factors contributing to human welfare are captured by it. Consequently, market prices and economic indicators based on them, such as national income and cost–benefit analyses, are misleading and thus must be corrected. The factor for which correction is most urgently needed is the environment.
>
> Tinbergen, J. and Hueting, R. (1992) GNP and market price: signals for sustainable economic success that mask environmental destruction, Chapter 4 (pp. 52–62) of *Population, Technology and Society*, edited by R. E. Goodland, H. E. Daly and S. El Serafy (Island Press, Washington) 156 pp.

Though the detail of the new field of ecological economics is a matter for professional economists, it seems clear that, at the very least, a reorientation and adaptation of traditional neo-classical approaches are required.

## Long-term Environmental Management

The past decade has seen an extensive debate on the basic strategy to be adopted in regard to management of the global environment. There has been much discussion of three principles: (a) the 'keep-options-open' principle, (b) the 'no-regrets-principle' and, (c) the 'precautionary principle'.

### *The Keep-Options-Open Principle*

Because the future is uncertain, it is wise not to foreclose options. For example, if the most likely response of an inland lake to global climate warming is a drop in lake levels, but there is a possibility of greater year-to-year swings in lake levels than at present, the hydrologist would be wise to consider the effects of both possible 'futures' on shoreline development. Unfortunately, some advocates of the 'keep-options-open' principle have in their opposition to the 'precautionary principle' (see below) sometimes sailed dangerously close to a 'do-nothing' approach.

### *The No-Regrets Principle*

The 'no-regrets' principle, under which action is taken which can be justified on other grounds, as well as those relating to the issue being considered, makes obvious sense for decision-makers in both the public and the private sectors. Examples of this approach are (a) promotion of increased energy efficiency to reduce $CO_2$ emissions, which at the same time reduces $SO_2$ and $NO_x$ emissions and conserves coal/oil/gas supplies for use by future generations; (b) the extension of product cycles to recycle wastes or to use them as raw material for further profitable processes; and (c) innovations in energy conversion which offer a major opportunity for reducing environmental damage without affecting the benefits of activities requiring high energy usages.

Policies based on the above approach can be of benefit, even in the relatively short term, for both developed and developing countries. Michael Porter, a professor at the Harvard Business School, concluded as a result of his study of competitive advantage between exporting countries that rigorous environmental standards do not inhibit business opportunities.

> I found that the nations with the most rigorous requirements often lead in exports of affected products... . The strongest proof that environmental protection does not hamper competitiveness is the economic performance of the nations with the strictest laws.

*The Precautionary Principle*

The 'precautionary principle' – that lack of scientific certainty should not be used as a reason for postponing measures to prevent unacceptable environmental degradation – has been widely debated over the past decade but there is now general consensus that the principle provides a sound basis for long-term environmental management. (Think long-term; act now!) It may be surprising to note that the principle was accepted by leaders of the business community many years before it was accepted by political leaders. The precautionary principle was agreed at the first World Industry Conference on Environmental Management in 1984, and the principle was strongly reaffirmed at the second such Conference in 1991. The Final Declaration of the latter Conference includes the following:

> We subscribe to a rational and precautionary approach to anticipating and preventing the causes of serious or irreversible environmental degradation consistent with scientific and technical understanding and economic use of resources.

The World Business Council for Sustainable Development also endorsed the principle (in 1992):

> Corporate leaders are used to examining certain negative trends, making decisions, and then taking action, adjusting and incurring costs to prevent damage. Insurance is just one example. There are costs involved, but these are costs that the rational are willing to bear and costs that the responsible do not regret, even if things turn out not to have been as bad as they once seemed. We can hope for the best but the 'precautionary principle' remains the best practice in business as well as in other aspects of life.

The World Business Council goes on to point out that risk and uncertainty are usually accompanied by new opportunities which are the life blood of successful business activity.

The political community has reached the same position as the business community. The precautionary principle was agreed to at the 1989 Paris summit of the G7 and was endorsed and strengthened in the following year by the UN Economic Commission for Europe in a Ministerial Declaration which stated:

> In order to achieve sustainable development, policies must be based on the precautionary principle. Environmental measures must anticipate, prevent and attack the causes of environmental degradation. Where there are threats of serious or irreversible damage, lack of full scientific certainty should not be used as a reason for postponing measures to prevent environmental degradation.

A similar view was propounded in November 1990 in the Ministerial Declaration of the Second World Climate Conference. Finally in June 1992, the precautionary principle was included as Principle 15 of the Rio Declaration on Environment and Development adopted by 179 countries.

While the precautionary principle has been agreed by both business leaders and political leaders, its implementation is not necessarily straightforward, either at an industry or a national level, partly because there are no agreed guidelines on how to operationalize the principle; see, e.g., O'Riordan and Cameron (1994). Some methods for introducing risk into environmental assessment are described in Appendix 4. By way of introduction here, it should be emphasized that many of the proposals for new environmental paradigms (see Box 3, for example) can be fully supported as responses to current conditions – but they are inadequate to meet the challenges that will be thrown up if unprecedented environmental, demographic and socioeconomic changes occur in the 21st century. (Consider, for example, the demographic projections for the year 2050 by H. N. Le Houérou for the southern side of the Mediterranean watershed, which range from 850 million (optimistic estimate) to 1950 million (pessimistic estimate), as compared with the current population of only 290 million! If such changes were to occur, the advice contained in Box 3 would do little to alleviate the environmental degradation that would inevitably occur.)

**Box 3. Ten Related Principles Defining a Proposed New Paradigm for Environmental Governance**

1. Stimulate new investment in industry and agriculture to capture the opportunities inherent in environmentally advantaged technologies.
2. Promote upstream solutions that anticipate and 'design out' potential environmental problems.
3. Use economic incentives and market-based mechanisms such as tradeable permits and pollution charges.
4. Integrate domestic and international environmental concerns.
5. Supplement conventional environmental approaches ('environmental laws') with high payoff initiatives to agriculture, energy and transportation, and internationally, with strengthened programmes for population and development assistance.
6. Promote voluntary and non-regulatory initiatives by both producers and consumers.
7. Replace confrontation with a new spirit of collaboration among government, industry and environmentalists.
8. Open the door to experimentation and flexibility in environmental regulation.
9. Launch a period of environmental law reform involving rationalizing and simplifying the vast body of rules and regulations that have accumulated.
10. Develop a broader consensus on the priority problems and allocate resources accordingly.

J. G. Speth, R. E. Train and D. M. Costle, *WRI Issues and Ideas*, November 1992 (WRI, Washington, DC) 4 pp.

## Involving the Public

Stakeholders concerned with global environmental problems have been increasingly interacting with one another over the past decade. That this trend should be accelerated in the next decade has been recommended by various international bodies, and suggestions have been offered on further steps that might be taken towards dialogue and partnership. For example, ICSU has recommended that 'increased efforts should be made by scientists to communicate with policy makers, the media, and the general public about the implications for society of the results of their research', and that 'continued efforts should be made to build bridges between leaders of science, business and industry, and consumers'.

In practice, communication amongst these groups is distinctly more difficult than might have been anticipated. What appear to be common objectives often involve shades of emphasis that create difficulties – not by provoking disagreement but by giving an appearance of a degree of agreement beyond what actually exists. What appears to be a common language may be common only in vocabulary and grammar with somewhat different meanings for identical words. (The phrase *sustainable development*, for example, has no equivalent in the Polish and Russian languages.)

Thus the message received may differ from the message sent, leading to later misunderstandings and loss of mutual trust. Only a disciplined willingness to listen, and a patient search for unambiguous consensus can avoid these pitfalls.

The relationship between science and society is a complex one. An interesting illustration of the nature of the problem has been given in an article by Gary Taubes in *Science* (11 June 1994). In the context of stratospheric ozone depletion, Taubes described the dilemma of atmospheric researchers on how best to convey to the public what is agreed to be highly probable and what is still uncertain in the context of the impact of CFCs on the ozone layer, and the resulting polarization in the public debate as to the most appropriate policy in regard to reducing emissions. How should a scientist respond when a self-styled 'expert' states at a public meeting that there is so much chlorine in sea spray that the contribution of chlorine in CFCs to stratospheric chlorine is trivial?

Spokesmen for the business community have also asserted the importance of communication with the public. The Declaration of the World Business Council for Sustainable Development emphasizes that new forms of cooperation amongst government, business and society are required to achieve the goal of sustainable development. The Declaration speaks of the approach needed in the following terms:

> The world is moving towards deregulation, private initiatives, and global markets. This requires corporations to assume more social, economic and environmental responsibility in defining their roles. We must expand our concept of those who have a stake in our operations to include not only employees and shareholders but also suppliers, customers, neighbours, citizens and others. Appropriate communication with these stakeholders will help us to refine continually our visions, strategies and actions.

The Schmidheiny book cites a number of examples of public pressures of different types (e.g., investor sensitivities or consumer pressure on environmental grounds) and of corporate reaction (e.g., use of advisory panels or the restructuring of management).

The Rotterdam Declaration of the Second World Conference on Environmental Management includes the following paragraphs:

> We will, as called for in the World Business Charter for Sustainable Development, foster openness and dialogue with employees and the public, anticipating and responding to their concern about the potential hazards and impacts of operations, wastes or services, including those of transboundary or global significance.
>
> We will periodically provide information to shareholders, employees, government and the public, and recommend specifically that companies include a statement by the Board of Directors or the Chief Executive Officer in their annual report, or in a special environment report, on the environmental status of the company. Some companies may choose to reinforce this by independent certification.
>
> We see the need to strengthen new relationships amongst governments, business, other stakeholders and the public at large. We encourage the International Chamber of Commerce to arrange Regional Environmental Roundtables to bring together regional leaders from industry, governments, trade unions, science and the NGO community to explore specific policies and initiatives.

Intergovernmental Conferences have similarly stressed the need to involve the public. Thus the Ministerial Declaration of the Second World Climate Conference includes the following:

> We believe that a well-informed public is essential for addressing and coping with as complex an issue as climate change, and the resultant sea-level rise, and urge countries, in particular, to promote the participation at the national, and when appropriate, regional level of all sections of the population in addressing climate change issues and developing appropriate responses. We also urge relevant United Nations organizations and programmes to disseminate relevant information with a view to encouraging as wide a participation as possible.

As another example, the International Conference on Water and the Environment, held in Dublin in January 1992, included the following paragraph in its Conference Declaration:

> Water development and management should be based on a participatory approach involving users, planners and policy-makers at all levels. The participatory approach involves raising awareness of the importance of water among policy-makers and the general public. It means that decisions are taken at the lowest appropriate level, with full public consultation and involvement of users in the planning and implementation of water projects.

The above principle is relevant to many other global environmental issues.

Finally, at the Earth Summit in Rio in June 1992, Principle 10 of the 27 principles comprising the Declaration dealt with public participation and openness:

> Environmental issues are best handled with the participation of all concerned citizens, at the relevant level. At the national level, each individual shall have appropriate access to information concerning the environment that is held by public authorities, including information on hazardous materials and activities in their communities, and the opportunity to participate in decision-making processes. States shall facilitate and encourage public awareness and participation by making information widely available. Effective access to judicial and administrative proceedings, including redress and remedy, shall be provided.

Public awareness and public participation are thus recognized as a vital element in the process of applying the results of scientific research to the solution of key problems relating to the environment.

## Selected References


Arrow, K., Bolin, B., Costanza, R., Dasgupta, P., Folke, C., Holling, C. S., Jansson, B., Levin, S., Mäler, K.-G., Perrings, C. and Pimental, D. (1995) Economic growth, carrying capacity and the environment, *Science*, **268**, 520–521.

Bernthal, F. D. (ed.), 1990, *Climate Change: The Response Strategies*. Intergovernmental Panel on Climate Change. WMO/UNEP, Geneva, 270 pp.

Costanza, R. (1991) *Ecological Economics: The Science and Management of Sustainability*. Columbia University Press, New York, 435 pp.

Daly, H. E. (1992) From empty-world economics to full world economics. Chapter 2 (pp. 23–37) of *Population, Technology and Society*, edited by R. E. Goodland, H. E. Daly and S. El Seraf. Island Press, Washington, 156 pp.

Daly, H. E. and Cobb, J. B. (1989) *For the Common Good: Redirecting the Economy towards Community, the Environment and a Sustainable Future*, Beacon Press, Boston.

Dooge, J. C. I., Goodman, G. T., la Rivière, J. W. M., Marton-Lafevre, J., O'Riordan, T., Praderie, F. (1992) *An Agenda of Science for Environment and Development into the 21st Century*. Cambridge University Press, Cambridge, 331 pp.

ECE (1990), Declaration of Ministerial Meeting. UN Economic Commission for Europe, Bergen.

FIDIC (1990) FIDIC on the Environment. Special issue of *Independent Consulting Engineer. J. Int. Federation of Consulting Engineers*, Autumn 1990, Lausanne, Switzerland.

FIDIC (1991) Harmonisation between Man and the environment. Independent Consulting Engineers, International Federation of Consulting Engineers, Lausanne, Switzerland.

G7 (1989) Economic Declaration Summit of the Arch, Paris, 16 July 1989.

Houghton, J. T., Jenkins, G. J. and Euphraums, J. J. (eds) (1990) *The Scientific Assessment*. Cambridge University Press, Cambridge, 364 pp.

ICWE (1992) Report of International Conference on Water and the Environment. Dublin, January 1992. WMO, Geneva.

Jäger, J. and Ferguson, H. L. (1991) *Climate Change: Science, Impacts and Policy*. Proceedings of the Second World Climate Conference. Cambridge University Press, Cambridge, 578 pp.

Le Houérou, H. N. (1993) La Méditerranée en l'an 2050. *Peuples Méditerranéens*, **62–63**, 29–63.

Mill, J. S. (1848) *Principles of Political Economy*.

O'Riordan, T. and Cameron, J. (1994) *Interpreting the Precautionary Principle*, Earthscan, London, UK, 315 pp.

Porter, M. (1991) Green Competitiveness. *New York Times*, 5 June 1991.

Schmidheiny, S. (1992) *Changing Course. A Global Business Perspective on Development and The Environment.* MIT Press, Cambridge, MA, 374 pp.

Taubes, G. (1993) The ozone backlash, *Science*, **260**, 1580–1583.

Tegart, W. J. McG., Sheldon, G. W. and Griffiths, D. C. (eds) (1990) *Climate Change: the IPCC Impacts Assessment.* Australian Government Publishing Service, Canberra, 290 pp.

UN (1992) Agenda 21. UN Publication – Sales No. E.93.1.11. 294 pp.

WCED (1987) *Our Common Future.* Report of World Commission on Environment and Development. Oxford University Press, Oxford.

WICEM-II (1991) Final Declaration of Second World Industry Conference on Environmental Management, 10–12 April 1991.

Willums, J.-O. (1990) (ed.) *The Greening of Enterprise. Business leaders speak out on environmental issues.* International Chamber of Commerce, Paris, 268 pp.

Willums, J.-O. and Golüke, U. (1992) *From Ideas to Action. Business and Sustainable Development.* International Chamber of Commerce, Paris, 351 pp.

WMO (1979) *Proceedings of the World Climate Conference.* WMO Publication No. 537. Geneva, 791 pp.

WMO (1986) *Report of the International Conference on the Assessment of the Role of Carbon Dioxide and of other Greenhouse Gases in Climate Variations and Associated Impacts*, WMO Publication No. 661, Geneva, 78 pp.

# 6. Intergovernmental Responses

P. TIMMERMAN and R. E. MUNN
*Institute for Environmental Studies, University of Toronto, Toronto, Canada*

## The Changing Role of Science in Global Policy

The rise of global science in the last 30 years has paralleled a shift in the role of science and scientists from pure advisors of how to put the cornucopia of scientific and technological advance to work, to a more ambivalent role, again foreshadowed by the concerns of scientists about the control of atomic devices. This new role is one in which, among other things, the long-term implications or complex outcomes of certain activities need to be considered. This 'early warning' role was – and is – difficult and ambivalent for many scientists, since uncertainty is pivotal towards scientific endeavour. Furthermore, since public involvement and understanding are now seen as essential to creating a climate of support for science, scientists have had to enter into public fora, and learn to redescribe their often complex activities for a wider audience.

More importantly, in areas such as genetic engineering and various forms of environmental toxicology, science and technology have begun setting the political agenda, and not merely responding to it. The most recent examples of this were provided by the Rio Earth Summit in 1992, where the largest part of the agenda was based on science – though the results had to be expressed in terms of politics and economics. Climatic change, biodiversity, forest management – these are issues where changes in scientific understanding have forced the pace of debate.

Without much exaggeration one could say that in Rio the countries of the world took note of the fact that science has produced evidence that the Earth System is endangered if human activities continue at the present pace, and that henceforth international policymakers must use scientific research and its results as an indispensible ingredient of national, regional and global environmental policy.

Increasingly science provides the 'ground-rules' for such issues, and the basis for many kinds of policy decisions. This is, of course, particularly true in environmental issues, where science is foundational. More problematically, in these areas the relationship between economic, political, and scientific activity is becoming more entangled, if not blurred.

For instance, the negotiations over the climate change convention resulted in a commitment to transfer technology as well as money to developing countries, to reduce as

*R. E. Munn, J. W. M. la Rivière and N. van Lookeren Campagne (eds), Policy Making in an Era of Global Environmental Change, 113–126.* 

much as possible the impacts of their need for continued growth over the near future. The moneys are transferred (in part) through the Global Environment Facility (GEF), an arm of the World Bank and UNEP. But project acceptability for funding through this process is in large part dependent on scientific evaluation. One problem is that the levels of uncertainty as to what would constitute a significant contribution to a carbon 'source' or 'sink' are still substantial; a second is that there remains conflict over whether only 'direct' rather than 'indirect' contributions to reducing source strengths or enhancing sinks should be counted.

It is interesting that as an adjunct to a purely scientific issue there are questions of equity involved. Developing countries find themselves necessarily beholden to developed country science and scientists for much of their information; and in certain negotiations this puts them at a disadvantage. There are also serious questions about how these countries will be able to conduct the analyses and monitoring required for adherence to specific commitments. This indicates that more and more reliance will be necessarily be placed in the near future on the international scientific community and its institutions to act in aid of developing country science, and as a neutral arbiter of international scientific accords.

## The Need for International Assessments of Global Environmental Issues

Principle 17 of the 1992 Rio Declaration states that:

> *Environmental impact assessment* (EIA), as a national instrument, shall be undertaken for proposed activities that are likely to have a significant adverse impact on the environment and are subject to a decision of a competent national authority.

This principle has already been adopted in many countries and by such bodies as the World Bank and the EU. In addition, where potential impacts of a development transcend national boundaries (e.g., pollution in large international rivers; acidic deposition; stratospheric ozone depletion; climate warming), environmental assessments are increasingly being undertaken. Beginning with the United States Department of Transport assessment of the impacts of supersonic aircraft on the stratosphere in the early 1970s, many such EIAs have been carried out.

International EIAs are often subject to the same faults as national ones (e.g., difficulty in dealing with long-term uncertainty, and identification and fair treatment of both winners and losers). In addition, international EIAs sometimes suffer from the problems of differing environmental standards and differing socioeconomic/environmental priorities in different countries. Indeed, some scientists argue that international EIAs should include consideration of the ramification of environmental change on social systems (Kotlyakov *et al.*, 1988).

Scientists have often treated international EIAs as purely scientific assessments written by and for scientists, the goal being to provide a consensus on the nature of the issue

**Some Examples of International Environmental Assessments**

| | |
|---|---|
| 1972: | The Swedish Case Study on Acidic Deposition prepared for the UN Conference on the Human Environment |
| 1974: | The assessment of SST stratospheric ozone depletion by the US Department of Transport |
| 1975–1979: | The assessment of acidic deposition in Europe leading to the ECE Convention on Long-Range Transboundary Air Pollution |
| 1978–1987: | The assessment of stratospheric ozone depletion leading to the 1987 Montreal Protocol on Substances that Deplete the Ozone Layer |
| 1990: | The assessment of acidic deposition by the US National Air Pollution Assessment Program (NAPAP) (a North American-scale assessment) |
| 1990: | The climate change assessment by IPCC (Intergovernmental Panel on Climate Change) |
| 1990–1995: | The second climate change assessment by IPCC |
| 1995: | The UNEP global biodiversity assessment |

being examined and on research priorities. In the case of NAPAP (the U.S. National Air Pollution Assessment Program), a North American-scale EIA, for example, there was criticism that the rather lengthy assessment produced in 1990 was not focused on an examination of policy options, which was the main reason why the US Congress had funded it. However, there has been gradual improvement in the usefulness of international assessments as a policy-forcing document. For example, the 1990 Policy-makers' Summary of the Scientific Assessment of Climate Change of IPCC (Intergovernmental Panel on Climate Change) was a model of how such EIAs should be presented, with the main results and recommendations given under the following headings:

*We are certain of the following: ...*
*We calculate with confidence that: ...*
*Based on current models, we predict: ...*
*Our judgement is that: ...*
*To improve our predictive capability, we need: ...*

## An Example: the Climate Change Issue

The possibility of climate warming due to rising concentrations of greenhouse gases has been a concern of many scientists since the 1970s. In 1979 a scientific assessment was undertaken for the US Academy of Science by a panel of eight specialists. In general, the Panel agreed with existing estimates that a doubling of greenhouse gases would cause a rise of about 3 deg C in world temperature but also stressed the uncertainties in this estimate. Subsequently, a scientific assessment was begun by SCOPE of ICSU in 1982, and mainly carried out at the International Meteorological Institute in Stockholm, financially supported by UNEP and WMO. The results of the assessment were presented at an international conference in Villach, Austria in 1985. The outcome of the conference was an appeal to politicians for action to curb greenhouse gases. Since Villach, 'climate warming' has been a public issue. In 1987 an intervention at the UN General Assembly by the representative of the Maldives focused attention on the impacts that a possible rise in sea level would have on low-lying islands. This placed 'climate change' on the UN agenda, and led to the creation in 1988 of IPCC (Intergovernmental Panel on Climate Change) by UNEP and WMO. At about the same time (1988) the International Conference on the Changing Atmosphere (Toronto) recommended that governments and industry:

> Reduce $CO_2$ emissions by approximately 20% of 1988 levels by the year 2005 as an initial global goal.

For industrial and government bodies whose mission had been to reduce emissions of conventional air pollutants (oxides of sulphur and nitrogen; particulates; toxics; etc.), this was quite a change in direction in technology and regulatory development. Here was a trace gas whose atmospheric concentrations had admittedly risen over the last 40 years but with as yet no statistical evidence of an accompanying rise in world temperature. Furthermore, the gas was perfectly harmless to people, enhanced growth of forests and agricultural crops, and was found in abundance under natural conditions all over the world.

In 1990 at the Second World Climate Conference, the scientific programme was followed by a Ministerial Session attended by Ministers from 137 countries. It produced a Ministerial Declaration which called for *inter alia* a Framework Convention on Climate Change. In that same year, the IPCC completed its first climate change assessment.

In retrospect it is remarkable that so soon afterwards (1992), a climate convention was signed by 156 countries at Rio at the United Nations Conference on Environment and Development (UNCED).

The Chairman of the IPCC, Bert Bolin of Sweden, has given his personal view of the lessons learned from four years of IPCC activity*:

(1) IPCC assessment is a continuing process of collaboration between scientists and politicians, a mutual learning experience. Scientists as well as politicians need to recognize their different roles. The former must protect their scientific integrity but also must respect the role of politicians. Scientists must also be viewed as honest representatives of their scientific colleagues, to ensure that the assessment process maintains its credibility.

(2) Scientists need to inform politicians in a simple manner that can be readily understood, but the message must always be scientifically exact. In reality, little of what scientists know is politically interesting or even understandable. Politicians are seldom scientists. It is difficult to sift objectively through the available sci-

* Summarized with permission from a paper by Bert Bolin in *Ambio*, Volume 23, February 1994, pp. 25–29.

entific information and extract what is politically relevant. For that reason, IPCC has published its reports in two parts: a scholarly scientific overview of about 300 pages; and a policy-makers' summary of about 25 pages.

(3) Although the IPCC was given the task of assessing and presenting available knowledge objectively, there is danger that political value judgements may also penetrate into the IPCC process. Although such a tendency must be guarded against, it probably cannot be completely avoided.

(4) Agreements on preventive or adaptive measures by society should be based on a combination of factual scientific information as provided by the IPCC and on value judgements. It is not the task of the IPCC to recommend actions, but rather alternative possibilities, pointing out their consequences.

The IPCC climate change assessment is a good example of the type of international environmental assessment that should be carried out whenever there is serious concern by a majority of informed scientists concerning an issue that may still be unfolding. Some other issues that have been assessed or should be assessed include:

- stratospheric ozone depletion
- acidic deposition
- the efforts of international trade agreements on the environment
- transboundary shipments of toxic chemicals
- biodiversity
- freshwater resources
- ocean pollution.

Some of the desirable characteristics of such assessments include the following:

(1) The assessments should be based on solid science, including appropriate models and field data.

(2) The assessments should be policy-driven, exploring options, alternatives and uncertainties.

(3) The assessments might include recommendations on research priorities but should not make policy proposals (e.g., to reduce emissions of greenhouse gases).

(4) The issues should be long-term and large-scale (affecting at least several countries).

(5) The assessments might include economic analyses, designed to answer well formulated questions, e.g., how much will it cost to reduce greenhouse gas emissions by $x\%$ vs. how much will it cost if these emissions are not reduced? These questions should be framed by policy analysts in collaboration with environmental specialists in the area concerned.

(6) Think long-term: act now! The inertia of the socioeconomic structure of the world and of individual countries is large, particularly if major investments have

been made in existing infrastructure. Major changes in energy supply systems, for example, would take decades to accomplish. Here it is important to emphasize that capital costs will increase, the longer the delay in initiating mitigation and adaptation strategies.

**The Net Costs of Mitigating Climate Changes**

Costs for a major mitigation programme, socially as well as economically acceptable, and for developing as well as developed countries, should be compared with the damages caused by climate change. Some economists believe that mitigation costs would be about 1–3% of the GNP of developed countries. A gradually increasing effort during a decade with the aim of resolving the climate change issue would therefore require an annual increase of available resources of about 0.2% of GNP. How does this compare with the damages that might ensue if climate warming were not checked?

In view of the ambition of both developed and developing countries to increase GNPs by at least 2% per year, greenhouse gas reduction of the required magnitude is hardly achievable.

Bert Bolin, the Huygens Lecture, November 1993

## Using Scientific Information to Establish Global Environmental Policy

In 1989, *Scientific American* published a special issue entitled 'Managing the Planet'.[†] Our new-found ability as a species to affect planetary physical systems – as exemplified by the prospect of global warming – does not mean, of course, that we are within sight of being able to manage the planet effectively or appropriately. We are hardly capable of managing ourselves. Nevertheless, the emergence of a global perspective, and our increasing capacity through satellites and computers to monitor and model global systems, does raise some fundamental issues about our planetary goals and purposes. The scientific and policy community, by and large, tends to assume that increased management of the earth and its resources is an inevitable task, whether we like it or not.

If access to information is a central defining characteristic of the powerful, then access to global scientific information becomes a source of potential conflict. On the one hand, the growing availability of satellite information allows nations to map and explore unexploited regions, to monitor environmental disturbances, and to predict impacts of development. On the other hand, much of the most valuable information is becoming privatized, and inaccessible to the public, including the wider scientific community.

As already mentioned, the differential access to scientific information is a chronic difficulty for developing countries, both for their official science advisors, and for the

[†] A question-mark following the word 'planet' was lost during the production phase of publication of the journal.

NGOs who seek to appraise the decisions being made. Here again, the results are ambivalent: access to certain kinds of information, thanks to e-mail and fax machines, is much broader than ever before; yet simultaneously, requirements for more specific information have never been greater.

Mere access to data is not enough. Every country also requires specialists to ask the right questions of the plethora of existing data banks, to understand the answers, and to apply the results.

So we need to ask what the global information being gathered is to be used for, who is to have access to it, and into what overall managerial framework is it being integrated. It is fairly clear, for example, that the management of environmental resources such as fisheries has been a series of catastrophes. This stems from a range of problems, some connected to managerial models that are quite inappropriate to ecosystems (e.g., promoting maximum 'sustainable' yield). In order to extract resources from nature, human beings have consistently sought to simplify, rearrange, and disturb ecosystems. Our increasing encroachment on planetary ecosystems also raises ethical questions about whether there will soon be room on the planet for creatures that do not serve human purposes however defined. Some people argue that to sustain a human population of 10–20 billion, we shall be forced to turn the planet into a factory farm. This will be planetary management with a vengeance.

Issues of 'sustainable development' and the long-term future of the planet require us to consider the new burden of responsibility placed on all members of the human community, but particularly on the international scientific community to advance understanding of the Earth System and to ensure open access to its results. Also, the widest possible debate is required about the implications of our emerging understanding of global environmental change.

## Strategies for Dealing with Global Environmental Change

### *Range of Strategies Available*

Strategies establish the fundamental orientation of responses within which more specific tactics are deployed. Global environmental change is characterized by great complexity and uncertainty, especially in linking local causes to global effects, and global causes to local effects. As a result, the strategic orientation is critical: decision-makers of all kinds, when faced with the prospect of global change, need to be able to interpret all kinds of disparate – and sometimes contradictory – information in such a way that the overall 'narrative' continues to make sense. For example, in the case of climate warming, with its spatial and temporal differences among countries, the onset of warming might not be smoothly linear, but may be characterized by increased climate variability tending in the direction of future warming or a higher frequency of occurrence of extreme events. This variability will make it hard to keep consistent support for action, unless an underlying strategic consensus is sustained (subject, of course, to revision).

Given the uncertainties, and the need to take a long-term perspective, it is interesting that on a number of issues, society has opted to begin by establishing overall principles or working guidelines, rather than focusing immediately on detailed responses.

Response strategies generally fall into the categories of:

(1) prevention – acting to prevent the change from occurring;
(2) mitigation – acting to modify the change, or to mitigate the worst of its possible consequences;
(3) adjustment – accepting the change at least temporarily, and adjusting one's actions accordingly;
(4) adaptation – accepting the change and permanently adapting one's actions accordingly.

These categories tend to overlap in the real world, but there are often advantages in focusing on one main strategy. For example, again in the climate change case, it is fairly clear that we are committed already to some climate warming, but political attention has been focused on prevention as a strategy. This has been due to a working assumption that opting for a strategy of adjustment or adaptation would unfortunately signal that the global community has given up. In the related case of ozone depletion, however, a strategy of prevention was adopted to send a clear message that CFCs would have to be phased out as soon as possible.

A fifth strategic category is 'wait and see'. Some economists have argued that it would be more appropriate to wait as long as possible before responding to large-scale global changes, since future expenditures are likely to be better spent, and global society will have more resources to mobilize. Obviously also, funding becomes more easily available when disasters make headlines. This approach is comparable to carrying out a real-world experiment that might have disastrous consequences.

Others argue that 'hedging our bets', as in insurance, is a more appropriate response to long-term uncertainties. This prudential strategy depends on a recognition of what is at stake in gambling on the global future. Related to this strategy is the 'no-regrets' policy, which essentially suggests doing what we ought to be doing anyway, e.g., energy conservation, stopping deforestation, etc.

Little attention has been paid to what we might call 'emergency' strategies in the case of, for instance, a runaway greenhouse effect, or the collapse of some major ecosystem such as the boreal forest. These have only a small probability of occurrence in our current models, but they are not impossible. It is clear that one of the best reasons to take precautionary measures is to avoid emergency responses, which will tend to require forcible and blunt measures, triage and abandonment, which cannot help but undermine important social goods.

Certainly, whatever the strategy, investment in more understanding, i.e., more scientific research and monitoring, is essential. And it is acceptance of the precautionary

| Examples of 'No-regrets' Climate-warming Policies | | |
|---|---|---|
| Policy | Effect on greenhouse gases | Other beneficial effects |
| Tree planting | Reduced $CO_2$ concentrations | • Improved microclimate<br>• Reduced soil erosion<br>• Reduced seasonal peak river flows |
| Energy conservation | Reduced $CO_2$ emissions | Conservation of non-renewable resources for future generations |
| Energy efficiency | Reduced $CO_2$ emissions | Conservation of non-renewable resources for future generations |
| CFC emission control | Reduced CFC emissions | • Reduced stratospheric ozone depletion<br>• Reduced surface UV-B, skin cancer and blindness |
| Box 7, p. 77 in *The World Environment: 1972–1992.* UNEP/Chapman & Hall, London. | | |

and no-regrets principles that buy time for doing this research. This is an additional reason for applying these principles.

*Understanding, Forecasting and Detecting Global Change*

In many cases, concerns about global environmental change arise largely from the results of model simulations. These concerns may be strengthened if changes are already taking place in precursor variables (e.g., the steady rise in atmospheric $CO_2$ concentrations in recent years). Another possibility is that global change is already occurring and could be detected if suitably designed monitoring and assessment systems were in place (e.g., the stratospheric ozone hole over Antarctica could have been detected a few years earlier than it was if a serious analysis of satellite data had been undertaken).

In all three cases, a continuing programme of environmental and related socio-economic monitoring is required:

(1) To provide early warning of unexpected changes, (e.g., of the Antarctic stratospheric ozone hole).
(2) To develop and test models that predict change, and that forecast the impacts of such change on society (thus monitoring must include not only the variables of direct concern but also a whole range of related variables).
(3) To provide early confirmation of predicted changes, e.g., of climate warming.

One of the problems encountered in practice is that the enormous natural variability that exists in both geophysical and ecological time series makes trend detection difficult. To infer from observations, for example, that global stratospheric ozone has diminished by 10% in the last 20 years requires a major statistical assessment of data from ground stations, balloons and satellites. As another illustration, Stephen Schneider takes up the question of detection of global warming (see *Science*, **263** (1994), pp. 341–347). He concludes that the warming of half a degree C over the last hundred years 'is pretty much outside the range of most century-long natural fluctuations experienced in the past several thousand years'. Yet if society waits until 99% statistical confidence is achieved, the point-of-no-return for trend reversal will have long passed.

Global environmental monitoring systems must continually evolve in the light of new technology, new models and new questions that are posed. International bodies that coordinate global environmental monitoring programmes are well aware of these problems, and are constantly striving for higher performance standards. So for example, the IGBP includes studies on the design of monitoring systems to observe the extent and functioning of the world's vegetation cover while the WCRP has a number of pilot studies designed to improve our capability to monitor the climate system.

**Some Global Monitoring Systems**

*1. Planned*

GCOS: Global Climate Observing System (WMO/IOC/UNEP/ICSU) (in planning stage since 1992)

GOOS: Global Ocean Observing System (IOC/WMO/UNEP) (in planning stage since 1993)

GTOS: Global Terrestrial Observing System (proposed by UNESCO/FAO/ICSU/WMO/UNEP) (in planning stage since 1994)

*2. Operational*

WWW: World Weather Watch (WMO) founded 1963

IGOS: Integrated Global Observing System (WMO/IOC) founded 1967

GEMS: Global Environmental Monitoring System (UNEP/WMO) founded 1974

GAW: Global Atmospheric Watch (WMO) founded 1989

Global environmental monitoring is coordinated by UN bodies, particularly the WMO and UNEP. ICSU is frequently sought as a partner in these programmes, in particular in the planning stages, when scientific inputs are required in determining what should be measured, where and at what times. ICSU's role is to ensure scientific quality and objectivity. However, actual monitoring is carried out nationally. This leads inevitably to gaps, particularly over the open oceans and in the poorer countries of the world. Remotely-sensed data from satellites partially overcome the problem, but there is still a great need for 'on-the-ground' observing systems.

The best strategy for coping with global change is based on an ability to understand the processes causing global change, as well as an operational capability to detect global change. Although logically the first priority should be given to ensuring that these goals are achieved, the reality is that the world community looks upon the generation of improved understanding and the setting up and operation of associated observing systems as tasks that develop automatically and do not require a conscious international effort.

On the one hand, there is some truth in this. Long before the Rio Conference, the scientific community had mounted huge national and internationally coordinated global change research programmes, in many cases in collaboration with UN bodies. More recently, planning for three comprehensive global observing systems (GCOS, climate; GOOS, oceans; and GTOS, terrestrial) has been undertaken.

On the other hand, these activities suffer to a large extent from financial anaemia, particularly from what has been called the 'core-funding paradox'; the fact that although donors and governments insist that research and observation programmes be carefully planned and effectively administered and coordinated, the sponsoring agencies are reluctant to provide funds for these 'core activities'. This is because they are accustomed to funding only concrete projects.

There are, however, some encouraging although still weak signals that the situation is improving. A few years ago an informal International Group of Funding agencies (IGFA) was established, primarily to provide a platform for consultation and information exchange with respect to major international programmes. Within this framework, the matter of core funding is beginning to be discussed, as is also the case in the Mega-Science Forum of the OECD, which in 1993 recognized global change research as belonging to Mega-Science. Similarly in the case of climate change, the member states of the WMO appear to have begun to recognize this problem.

Meanwhile international programmes operate vigorously but sub-optimally; some projects are slowed down for lack of funds while others have to be postponed. In particular, there are insufficient funds available to launch GCOS, GOOS and GTOS as quickly as scientific prudence would dictate.

It is clear that research on global change and on human adaptation requires 'new and additional' funds, but there is still a long way to go before this is realized. Governments continue to give high priority to addressing issues of short-term concern, and are not willing to increase their contributions to international scientific efforts despite their signing of Agenda 21 at UNCED. In the present financial climate, priorities have to be set within three categories: increasing scientific understanding, improving forecasting abilities and taking precautionary action. This resembles the problem of deciding which leg of a tripod is the most important.

### *Actions to Follow a Scientific Assessment*

There are four kinds of actions that should follow the preparation of a scientific assessment of a global environmental problem:

(1) Actions to *reduce uncertainties* in the assessment (a scientific task) by sponsoring priority research and monitoring. The IGBP and the WCRP are examples.
(2) Actions to *slow down predicted rates of change*, e.g., through a reduction in the emissions of greenhouse gases and strengthening of the absorptive capacity of greenhouse gas sinks.
(3) Actions to *protect society from the harmful effects of change*, through structural engineering works, e.g., strengthening dykes, or increasing freshwater storage capacities.
(4) Actions to *promote adaptation strategies*, e.g., by moving populations away from low coastal areas, or by developing new genetically engineered strains of agricultural crops that can better withstand drought or heat. Adaptation is generally a national rather than an international activity, and thus is discussed in the following chapter. However, international bodies such as IPCC, WMO and UNEP contribute importantly to the development of national adaptation strategies through the preparation of guidelines, training manuals, etc.

Of course an integrated programme built upon all four courses of action is best. In this subsection we shall discuss action of type 2 – slowing down the change.

Using greenhouse gas warming as an example again, the 1990 international scientific assessment of the IPCC was a first step in achieving international political consensus on an action plan to reduce global emissions of greenhouse gases. Then at the Second World Climate Conference later in the same year, a Ministerial Declaration agreed to 'call for negotiations on a framework convention on climate change to begin without delay'. Subsequently, in December 1990, the General Assembly of the United Nations established an Intergovernmental Negotiating Committee to prepare 'an effective convention on climate change, containing appropriate commitments, and any related instruments as might be agreed upon'.

As mentioned earlier, the Climate Convention was signed by 156 countries at the United Nations 1992 Rio Conference on Environment and Development. A further step in the intergovernmental process is *ratification*, which requires the signatures of 50 countries before a Convention can come into force. This number has now been reached.

Some key paragraphs in the Climate Convention signed in Rio in 1992 are given in the box on page 125.

As pointed out by Cameron (1993),

> The consequences of the Convention pervade almost every aspect of modern economic life. Its development will be constrained by the extent to which governments will be prepared to qualify the unfettered access presently enjoyed by industrial societies to the limited resources of the global commons. It is, nevertheless, this pervasiveness of effects which marks out the convention as a radical step towards sustainable resource management.

Some key paragraphs in the Climate Convention are:

*Article 2: Objectives*
The ultimate objective of this Convention ... is to achieve ... stabilization of greenhouse gas concentrations in the atmosphere at a level that would prevent dangerous anthropogenic interferences with the climate system.

*Article 3: Principles*
The Parties should protect the climate system ... in accordance with their common but differentiated responsibilities ... the developed countries should take the lead in combating climate change... . The specific needs and special circumstances of developing countries ... should be given full consideration.

*Article 4: Commitments*
1. All parties ... shall:
(a) Develop, periodically update, publish and make available to the Conference of the Parties, in accordance with Article 12, national inventories of anthropogenic emissions by sources and removals by sinks of all greenhouse gases not controlled by the Montreal Protocol, using comparable methodologies to be agreed upon by the Conference of the Parties.

## Future Prospects

The emergence of a global framework of understanding divides policy-makers into those who envisage themselves as working within a 'global community' and those who continue to function within a pre-global framework, whether it be the nation-state (or indeed the city-state) or a belief in the infinite expansion of industrial growth. As humanity is forced to operate within more finite physical boundaries, new constraints as well as opportunities are being uncovered. On the other hand, we find that the nation state is embedded very deeply in international accords, and the freedom of nations to exploit their resources was strongly reaffirmed at Rio.

The struggle between the forces of global change and the need to protect and enhance local diversity will be the central strategic issue for the bulk of the next century in which the signals given by new information coming from scientific research will have an important role. It will be played out in part – and hopefully mediated – through international accords, institutions old and new (like the Commission on Sustainable Development of the UN and the NGO Earth Council), scientific agreement, and the activities of individuals, groups, and nation states. Whether common communities of concern will be strong enough to overcome the familiar patterns of national self-interest is, of course, a basic issue.

Before everything else, more knowledge is required of the Earth System, how it operates, what its limits are, and what opportunities exist for exploiting the changes expected.

## Selected References

Bolin, B. (1994) Science and policy making. *Ambio*, **23**, 25–29.

Bolin, B., Döös, B., Jaeger, J. and Warrick, R. A. (1986) *The Greenhouse Effect: Climate Change and Ecosystems*, SCOPE 29. John Wiley and Sons, Chichester, UK, 541 pp.

Cameron, J. (1993) The Climate Change Convention; how it was made and what it means. *Int. J. Env. Pollution*, **3**, 67–77.

Houghton, J. T., Jenkins, G. J. and Euphraums, J. J. (eds) (1992) *Climate Change 1992*. IPCC WG 1. Cambridge University Press, Cambridge, UK.

IPCC (1992) *Preliminary Guidelines for Assessing Impacts of Climate Change*. IPCC/WMO/UNEP, Geneva, Switzerland, 28 pp. (A second assessment is in preparation.)

IUCN/UNEP/WWF (1991) *Caring for the Earth: A Strategy for Sustainable Living*. IUCN, Gland, Switzerland.

Kotlyakov, V. M., Mather, J. R., Sdasyuk, G. V. and White, G. F. (1988) Global change: geographical approaches (a review). *Proc. Natl. Acad. Sci. USA*, **85**, 5986–5991.

Nilsson, A. (1992) *Greenhouse Earth*. John Wiley and Sons, Chichester, UK, 219 pp.

Roberts, L. (1991) Learning from an acid rain program. *Science*, **251**, 1302–1305.

Schneider, S. H. (1994) Detecting climatic change signals: are there any 'fingerprints'? *Science*, **263**, 341–347.

UN (1992) *Framework Convention on Climatic Change*. United Nations, New York.

# 7. Examples of Governmental Responses

P. TIMMERMAN and R. E. MUNN
*Institute for Environmental Studies, University of Toronto, Toronto, Canada*

## Range of Strategies Available to National Governments to Deal with Global Change

Historically, science played a minor role in national affairs, until perhaps the rise of the industrial state and the exigencies of modern warfare promoted science and its technological offshoots into becoming a major participant. The Second World War in particular dramatically showed how scientific and technological advances could affect, and even turn the tide of battles. The more obvious of these were advances in radar, airplanes, and the development of atomic weapons; but there were important roles played by, for example, increased sophistication in meteorological prediction. Symbolic of this new-found importance was the creation of the role of the science advisor to government, as well as Ministries of Science and Technology. This importance of science to government was reciprocated through vastly increased commitments of developed-country governments to funding science, either through internal government research or through universities.

The following strategies are available to national governments to deal with global change:

- *Knowledge-building strategies*: the promotion of research to improve understanding of global change is an essential ingredient of national response strategies.
- *Capacity building*: for developing countries, capacity building is an essential ingredient of their responses to global change. Many nations do not have sufficient specialists to assess the national implications of such change, and to develop appropriate response strategies.
- *Prevention/limitation strategies*: these are strategies designed to prevent, or to limit as far as possible, an expected environmental change. They do not refer to methods for limiting the damage of a change (by building sea walls, for example) but rather they are designed to slow the onset or diminish the magnitude of an expected change. *Example*: reduction of greenhouse gas emissions.
- *Adjustment strategies*: these are strategies designed to provide temporary adjustments to an expected environmental change. They are often seen as short-term,

*R. E. Munn, J. W. M. la Rivière and N. van Lookeren Campagne (eds), Policy Making in an Era of Global Environmental Change, 127–137.* 

locally restricted, experimental, reversible, not-yet permanently integrated additions to societal response mechanisms. *Example*: flood and drought relief.

- *Adaptation strategies*: these are strategies designed to create, sustain or entrench a permanent or long-term structural change as a response to an expected environmental change. *Example*: transfer of people from low-lying areas.
- *Restoration strategies*: these strategies are designed to restore a society, built infrastructure or an ecosystem to its original state and functioning – or equivalent – before an expected environmental change occurs. *Example*: afforestation.

There are obvious linkages and overlaps amongst these strategies, and governments tend to promote, either deliberately or haphazardly, a mixture of strategies. Nations have not only special priorities, but histories of response to risks which shape their selection of particular strategies. For example, the Netherlands has seen the potential for sea-level rise as not just a hazard, but also as an opportunity for application of its internationally recognized expertise in storm-surge hazard management. In addition, some nations, either because of their size or insularity, have preferred to focus primarily on internal responses to global change issues with local impacts, whereas others – the island state of the Maldives is perhaps the best known example – are almost wholly dependent on collective global efforts.

The extraordinary current debate taking place in many countries under different circumstances, is about the nature and implications of adopting either a single strategy or a mix of strategies, either separately or as part of the global response. For example, should efforts be put into entirely limiting climate change, or should nations resign themselves to a certain amount of warming, and determine how best to adapt? What is the proper ratio between limitation, adjustment and adaptation? Among the important elements of this debate, are questions about how well societies are currently adapted to the environment, how well are decision-making processes based on appropriate information, and what should be the role of government in encouraging, assisting or regulating change.

Generally speaking, it is in the interest of all nations to maximize their flexibility, their ability to respond – to keep their options open as long as possible, given the uncertainties – and to ensure that their institutions and peoples have the resources, determination and information required to cope. Among the supportive roles to be played by international organizations and the wider community of nations is to see that some of these capacities are available.

## Knowledge-Building Strategies

### *Relations Between National and International Global Change Research Programmes*

National governments in general are supportive of international global change research programmes such as IGBP and WCRP. Many of them also sponsor related research on:

(1) Methods to reduce their share of world-wide emissions of greenhouse gases, to reduce deforestation, etc.
(2) The effects of global change on their specific territories, and the development of appropriate policies and strategies in the light of their own stage of economic development.
(3) Related long-term issues such as sustainable development and carrying capacity.

However, the relationship between national and international global change programmes is complicated by a number of factors, including differing perceptions of the implications of global change and differing priorities and levels of available resources. Generally speaking, international programmes often act as 'frameworks' and catalysts for national programmes; and it is the movement back and forth between the two that accounts for much of the diversity as well as creativity across the global scene. It also makes the understanding and management of global research difficult. In particular, it is sometimes hard to prevent problems such as duplication of effort and inappropriate or wasted research. Continued coordination of national and international research is vital.

There is a history of international programmes and institutions, which successfully operate very often through 'national committees' and partnerships, thus helping to create and sustain a global infrastructure in many areas. There has been an evolution in this kind of cooperation, particularly in the movement of developing countries towards full and equal partnership in the management of collaborative research programmes.

Since countries have different timetables and areas of concern, there is no single model that applies to all projects and programmes. In some countries there is a long history of concern and expertise in an area, and a new international programme may have to consider ways of integrating already existing national research initiatives into its activities, just as national programmes may in turn partially feel the need to reorganize or 'add-on' an international component. In other areas, of course, the international programmes may be the initiator of interest, and nation states may have to be persuaded that collaboration in a particular topic area is in their own interest.

In general then, development of national programmes and participation in international programmes comes out of a complex mix of motives, including national self-interest, public concern, and occasionally desire to be a 'good global citizen'.

*Examples of Some National and Regional Global Change Research Programmes*

National global change research programmes have thus developed in part from international initiatives, and in part from internal institutions, very often national science bodies, such as National Academies and Royal Societies, which either oversee national research activities or occasionally act as a catalyst for ventures outside of familiar disciplinary research. Some national programmes have developed out of, or been connected to, earlier initiatives on sustainable planning for the nation. Several countries (e.g., France,

Germany and the United States) publish comprehensive illustrated annual reports describing their global change research activities. In what follows we present a selection of programmes from around the world. For additional information, see the Report of the Fourth Meeting of National IGBP Committees (IGBP, 1994).

The most advanced national programmes are in the OECD countries, with nations like the Netherlands, Germany, Great Britain and the United States having taken some early steps in the 1980s to build upon their extensive scientific bases in support of global physical research in the physical sciences. Each country has its own process for working towards national contributions to global change goals, including participation in international programmes such as IGBP. In the Netherlands, apart from such organizations as the National Research Committee for the IGBP, national response strategies are co-ordinated through the National Environment Policy Plans (NEPP) agreed upon each year, which incorporate what are called 'environmental outlooks' – assumptions about different global scenarios of social and economic development. In Germany, a lead was taken almost a decade ago by what are called 'study commissions' of the German Bundestag. As early as 1987, a Study Commission was established to consider and propose for implementation a national programme for protecting the earth's atmosphere. The acceptance of its extensive review of the situation led to the development of a substantial German presence in early IGBP work, and contributed to the establishment of the German Global Change Programme. At present, global change research in Germany is largely directed to IGBP and related WCRP Projects (Germany, 1993), with emphasis on:

(1) IGBP
  International Global Atmospheric Chemistry Project IGAC
  Joint Global Ocean Flux Study JGOFS
  Past Global Changes PAGES
  Land–Ocean Interactions in the Coastal Zone LOICZ
  Biospheric Aspects of the Hydrological Cycle BAHC
  Global Change and Terrestrial Ecosystems GCTE
  Contributions to central IGBP activities (Data and Information System DIS; Global Analysis, Interpretation and Modelling GAIM; System for Analysis, Research and Training START)

(2) Related WCRP
  Global Energy and Water Experiment GEWEX
  Stratospheric Processes and their Role in Climate SPARC

In Great Britain, approximately 120 million English pounds are spent annually by the government and its five Research Councils, as overseen by an Inter-Agency Committee on Global Environmental Change. British priorities have been strongly focused on basic science, on the assessment of potential human impacts, and on the socio-economic dimensions of global change.

Undoubtedly the largest and most significant national programme is that of the United States. In fiscal year 1994, the Federal budget for the US Global Change Research Program was 1.4439 billion dollars, subdivided as shown in the Box below.

**The US 1994 Federal Budget ($ million) for Global Change Research (US GCR, 1994)**

| | |
|---|---|
| Observing the Earth System: | |
| space-based | 579.8 |
| ground-based | 25.4 |
| Managing/archiving data/information | 277.4 |
| Understanding global change | 456.5 |
| Predicting global change | 49.5 |
| Evaluating the consequences | 38.6 |
| Developing tools to assess policies/options | 16.9 |
| Total | 1443.9 |

Examples of some recent accomplishments within the US Global Change Program are as follows (US GCR, 1994):

(1) Continuous global monitoring of stratospheric ozone revealed record low total ozone in 1993.

(2) Rates of deforestation in Brazil, as observed by Landsat, are now lower than previously believed.

(3) A national UV-B monitoring network has been established.

(4) Long-term retrospective global data sets, which include marine meteorological observations, upper air data, aerosol and ozone data, and baseline data on atmospheric trace constituents, have been developed and are being used by researchers worldwide.

(5) Studies on past climate from ice core data indicate that abrupt climate shifts occurred frequently in those regions over time scales on the order of decades or less over the last million years.

(6) A pilot project has been completed to train an initial group of international scientists in seasonal and interrannual climate modelling and forecasting, and in interpreting these forecasts for regional application by decision-makers for management of resources affected by El Niño/Southern Oscillation events in the Pacific Ocean.

(7) Regional-scale models of the impacts of global warming on reservoirs in the western United States have shown that the effects of temperature increases on aquatic ecosystems and on fisheries could be significant.

(8) Economic research has shown that the net costs of greenhouse gas emission reductions may be significantly lower than previously believed, when ancillary

benefits such as those due to simultaneous reductions in sulphur and particulates are included.

The Canadian Global Change Programme was established in 1989, and operates through the Royal Society of Canada. The bulk of the research is oriented towards the physical sciences, with special reference to areas of national expertise and interest, such as the arctic, and promoting a Canadian presence in existing or planned international experiments. The Canadian programme has 'human dimensions of global change' as a partner with IGBP in its overall planning structure.

**Objectives of the Canadian Global Change Programme**

- To develop a national framework for global change research
- To assist the policy sector with the creation of effective global change policies
- To assemble the appropriate research disciplines and organizations within the policy sector
- To foster interdisciplinary and interagency research and coordination
- To act as a clearing house for Canadian global change activities
- To act as the national liaison for international global change programmes
- To promote awareness of global change and of the opportunities arising from global change

As in many other countries, the French Global Change Programme is built around IGBP and WCRP, but with special emphasis on the particular scientific interests of the French research community (France, 1992), e.g., wind erosion in arid and semi-arid regions; analysis of the dynamics of West African savannah; tropospheric ozone and its precursors.

Japan has a National Committee for the IGBP, whose primary focus has been on physical science projects. The Japan Environment Agency established the Global Environment Research Fund in 1990. Its areas of study include participation in IGBP and the WCRP, the full range of physical problems associated with global change, including human impacts and planning policies to conserve the global environment. An important cluster of initiatives is associated with the National Institute for Environmental Studies in Tsukuba, which is, among other projects, developing an integrated scenario analysis for climate change in the Asia-Pacific region based on an end-use energy demand model. This includes models that are responsive to the introduction of new technologies, sensitive to the impacts of climate change on primary production, as well as including population and economic growth modules.

Global Change Research in Southeast Asia has been stimulated by the internal pressure of rapid economic growth in the region, and by the external pressures of international concern over such issues as rainforest depletion. Global and regional institutions that have made their presence felt include the World Bank, the Asian Development Bank, and the UN ECOSOC Committee for Asia and the Pacific. The ECOSOC Committee has been responsible for the initial State-of-the Environment reports for the region. The Asian Development Bank, for its part, is an active participant in the GEF (Global

Environmental Facility), and is preparing an investment programme based on its 'Development of Least Cost Greenhouse Gas Emission Reduction Plans in Asia' report, as well as country-by-country strategy reports. The Asian Pacific Network, just like the InterAmerican Institute for Global Change Research, works closely with the worldwide START network of IGBP, HDP and WCRP.

One country in the region which has pioneered in global change research is Thailand. The Thai response to global change is handled through its National Research Council, which operates the National Committee for the IGBP, founded in 1989. The first national workshop on the IGBP was held in Bangkok in 1990. By 1993, more than 13 institutes were involved in various IGBP-related projects. Thai interests have focused on research gaps such as the absence of studies on sea-level rise, methane emissions from paddy fields, and the special characteristics of a typhoon-prone area.

## Capacity-Building Strategies

A lack of qualified scientists within a country may be a bottleneck to achieving better understanding of global change, and to designing appropriate national response strategies, especially but not exclusively in developing countries.

The most important regional initiative in capacity building is the IGBP-sponsored START programme (the System for Analysis, Research and Training) which is designed to assist in training scientists and in establishing regional networks for undertaking global change research in the particular global change issues of the region, especially in areas that need substantial upgrading, particularly but not exclusively in developing countries. START helps ensure that governments have their own scientists (and not foreign consultants) to advise them in international negotiations for conventions dealing with global change, enabling them to participate on an equal footing. START is also important in assisting global research programmes to obtain worldwide coverage in their monitoring and field research programmes. Despite the valuable information received from satellites, there is still a serious lack of ground measurements in some parts of the world. Since its inception in 1991, START has been focusing on three priority areas: Equatorial South America, the Tropical Asian Monsoon Region, and Africa north of the equator.

In addition, two intergovernmental networks for global change research have been established: the InterAmerican Institute for Global Change Research (IAI), which is sponsored by most countries in Latin and North America; and the Asian Pacific Network (APN). The aims of both networks are very similar to those of START, and close collaboration is assured.

## Adaptive Strategies

Examining the situation in many countries around the world, very little serious consideration is being given to integrated development models, with a view to creating sustainable societies. Most national environmental policies are still characterized by piecemeal regulation and enforcement, emphasizing short-term solutions to immediate issues. In the same way, there is little ability to consider environmental changes – both degradation and improvement – as expressions of large-scale syndromes or social tendencies. As a result, national responses tend to be *ad hoc* and scattered. Population growth, inappropriate farming practices, excessive or unmanaged resource extraction, and so on, contribute to the stresses that force unwanted environmental change; but making the appropriate links and implementing appropriate adaptive strategies is always difficult.

Some strategies are even maladaptive, leading to responses in which a temporary adjustment to circumstances becomes a permanent obstacle to further change, often because it develops a constituency. A classic example of this is subsidies on farm production during hard times, which are difficult to remove when conditions improve. Prominent examples of these strategies can be found in the environmental hazards literature; see, for example, Burton *et al.* (1978). Sometimes these maladaptations are worse than doing nothing: a famous example in the hazards literature (especially in the United States) is the way in which temporary protections against hazards such as floods encourage in-migration to the newly protected areas, thus increasing the potential for greater catastrophes should a dam or dyke fail. It must always be kept in mind that preventing environmental damage is not necessarily the same as preventing unwanted environmental change.

A further complicating factor is that some environmental change is inevitable – the environment is changing all the time. What is of concern is unwanted environmental change or change occurring too fast for the effects to be assimilated, or in a specially vulnerable place. To attempt to freeze the environment at a particular moment is a strategy that has often proved to be a recipe for catastrophe, particularly as human societies tend to wish to freeze the environment at extremely high levels of productive yield. The collapse of fisheries in many parts of the world is partly due to this 'locking in' to the peak of an ecosystem's productivity, which is unsustainable.

Of course, one difficulty in responding to unwanted environmental change is that it is not necessarily unwanted by everyone: some constituencies thrive on this kind of change, just as disturbed environments open up new niches for opportunistic species. A deteriorating public water supply provides an opportunity for selling bottled water.

This differentiation of impacts from environmental change, and its social and political importance at the national, regional and local levels helps explain why there is so much pressure being put on modellers of global change to predict local impacts. Yet it has also been argued that the more clarity there is about potential winners and losers, the harder it will be to build up coalitions within and between nations to address the overall necessity for general limitation strategies.

Benign adaptation to environmental change comes in the forms of incremental adaptation (planned or inadvertent), and major planned adaptation. Since societies are adapting to environmental change all the time, the process whereby individual decisions or larger-scale social forces begin as adjustments to changes in the frequency of certain events, and become enshrined as adaptations is an historically familiar one. The cultures, costumes, shelters, etc. of the nations of the world were – until the advent of the global village – expressions of adaptation to local environments. These were more often than not incremental adjustments that over time became permanent adaptive fixtures.

Major planned adaptations are also historically familiar, especially when one considers the immense hydraulic works of ancient times. The ability to mobilize a whole society, which in the distant and recent past was usually the prerogative of tyrannies, is happily complicated in democratic societies by having to persuade and convince. Planning for change is, of course, controversial; some economists suggest that society will be most adaptive through the decisions of myriads of individuals, rather than through planning; others argue that there are market failures and non-economic issues that require an overall framework within which individual adjustment decisions are made. Some social scientists argue that there are moments in social development when adaptation requires a leap past existing structures, and this requires a high-level intervention. Support for such major adaptive decisions will be difficult in the absence of either a compelling vision or a catastrophe which will make people willingly sacrifice current adjustment benefits.

An example of what can be done to promote adaptive responses to environmental change is provided by Costa Rica, which simultaneously has one of the highest rates of deforestation in the world, and one of the largest percentages of land protected as natural parks and other natural reserves. In the 1980s, Costa Rica was the first Central American country to raise the environment to a cabinet-level position, as well as to initiate a conservation strategy along the lines proposed by the World Conservation Union (IUCN). In 1986, Costa Rica began a comprehensive national consultative planning process called the Costa Rican Conservation Strategy for Sustainable Development (ECODES) which prepared extensive sectoral reports, and conducted integrating analyses at the highest levels of government. This resulted in a framework for a national development strategy aimed at all sectors, including the legal system, public health, cultural resources and education. These were directly linked to the assessment of environmental change, and its potential impacts on national well-being. Such a process gives policy-makers and citizens ways of determining what environmental changes should be prevented or limited, and what changes should be encouraged.

An inventory of adaptation strategies relating to possible climate change, including increased variability, is given in Appendix 5 (Smit, 1993). The range of possibilities in the various economic sectors is quite enormous.

## Restoration Strategies

In an unchanging world, restoration ecology would be uncontroversial. In the rapidly changing world of the 21st century, however, the situation is not so clear cut. The dream of restoring an ecosystem to an earlier pristine state comes up against the hard fact that such a restoration would be taking place in the larger context of increasing human interference, managed ecosystems and global change. It is hard enough to stop the clock, let alone turn the hands back. This is not to suggest that some forms of restoration are not urgently required. In fact, when the term 'sustainable development' began to be commonly used in the 1980s, two Canadian ecologists coined the phrase 'sustainable re-development' to indicate that there were degraded ecosystems in all parts of the world that needed to be restored to at least some level of improved functioning (Regier and Baskerville, 1986). Suitable methods for undertaking re-development assessments are in general available, but it should be emphasized that it is not generally a good idea to return an ecosystem to its pristine state – this would reduce its resilience to change. This is an important question which has led ecologists into studies of non-equilibrium systems, which have the interesting property that they can flip rather suddenly and often unexpectedly into other systems. The phrase 'ecosystem integrity' has been coined to indicate the ability of an ecosystem to change/evolve successfully rather than flip or collapse in a changing environment: this capacity is connected to the self-organizing ability of the system. Although there are as yet few practical applications, this is a research topic of intense current interest. For additional information on ecosystem integrity, see Woodley *et al.* (1993) and references therein.

These ideas apply in principle also to socioeconomic, business and political systems such as city states and countries. However, the field is less advanced than it is in the case of ecological systems.

## Concluding Remarks

In addition to the range of strategies described above to deal with global change, there is the obvious approach of slowing down or even preventing global change. Doing this by edict (e.g., through the intergovernmental 'Framework Convention') will be difficult. However, business and industry can make significant contributions through the continuing development of engineering technologies (see Chapter 8). Motivating people to reduce their personal emissions of greenhouse gases, through for example reduced dependence on the automobile is probably the most intractable problem of all.

## Selected References

Burton, I., Kates, R. W. and White, G. F. (1978) *The Environment as Hazard.* Oxford University Press, Oxford.

Germany (1993) *IGBP Research in Germany.* IGBP Secretariat, Free University of Berlin, Carl-Heinrich-Weg 6-10, 1000 Berlin 41, Germany, 86 pp.

IGBP (1994) Report of the Fourth Meeting of National IGBP Committees, Bonn-Bad Godesberg, March 1994. IGBP Secretariat, Free University of Berlin, Carl-Heinrich-Weg 6-10, 1000 Berlin 41, Germany, 114 pp.

IGBP (1991) *Global Change System for Analysis, Research and Training (START)* (J. A. Eddy, T. F. Malone, J. J. McCarthey and T. Rosswall, eds.). IGBP Report No. 15, Royal Swedish Academy of Sciences, Stockholm.

IGBP (1992) Report from the START Regional Meeting for Southeast Asia. IGBP Report No. 22, Royal Swedish Academy of Sciences, Stockholm.

IUCN/UNEP/WWF (1991) *Caring for the Earth: A Strategy for Sustainable Living*. IUCN, Gland, Switzerland.

Regier, H. A. and Baskerville, G. L. (1986) Sustainable re-development of regional ecosystems degraded by exploitive development. In *Sustainable Development of the Biosphere* (W. C. Clark and R. E. Munn, eds) Cambridge University Press, Cambridge, pp. 75–101.

Rosemarin, A. and Svedin, V. (eds) (1988) Ecosystem redevelopment. *Ambio*, **17**, 83–152.

Smit, B. (ed.) (1993) *Adaptation to Climatic Variability and Change*. Report of the Task Force on Climate Adaptation, Atmospheric Environment Service, Downsview, Ont., Canada, 53 pp.

US GCR (1994) *Our Changing Planet: The US FY 1995 Global Change Research Program*. Nat. Sci. and Techn. Council, Washington, DC, 132 pp.

Woodley, S., Kay, J. J. and Francis, G. (1993) *Ecological Integrity and the Management of Ecosystems*. St Lucie Press, USA, 220 pp.

# 8. Examples of Responses by Business and Industry

N. VAN LOOKEREN CAMPAGNE
*Interviews by ir. Joost van Kasteren, Pica Press*



## Introductory Remarks

### *The Business Ecosystem*

The world business community can be compared with a large ecosystem such as a tropical forest or a reef containing many species and many ecological interdependencies (competition for nutrients, predators and prey, symbiotic relations, etc.). The business ecosystem too contains a diverse range of 'species', from multi-nationals to tiny entrepreneurs, and many interdependencies.

The 'Standard Industrial Classification' of business lists thousands of 'species', grouped in genera like agriculture, mining, construction, manufacturing, finance, trade, services and administration. Manufacturing alone contains 500 types of industry. The business ecosystem not only includes products, trades and services but also the consumers of these goods and services. People tend to forget that they are part of the business ecosystem.

In ecosystems, each 'species' and individual responds to changes in the environment in many different ways. Some species become extinct while others are able to adapt to a

*R. E. Munn, J. W. M. la Rivière and N. van Lookeren Campagne (eds), Policy Making in an Era of Global Environmental Change, 139–163.* 

changing environment. The same applies to business ecosystems. Individuals and species respond in their own ways to perceived future changes in the environment, changes that become apparent by the growing insight into the functioning of the Earth System.

An important difference between ecosystems and business systems is that the former are practically waste-free. The business/consumer-system still has a long way to go in recycling and resource management.

### *Business Spokespersons*

The business community is quite well organized through various sectoral, national and international associations. With respect to environmental issues, there are two main bodies that represent business (including industry) interests. The first one is the ICC, the International Chamber of Commerce in Paris. This entity has a first-category consultancy status at the United Nations. The second one is the World Business Council for Sustainable Development. This is a group of some 70 business people from all over the world. What makes them special is that they were asked by the Secretary General of UNCED in Rio de Janeiro 1992 to present their combined views to the Earth Summit. See Chapter 5 for more information.

In addition, there are two UN bodies that deal with business-related issues. These are UNIDO (the United Nations Industrial Development Organization) in Vienna and the UNEP (United Nations Environmental Program) Industry Office in Paris. Another organization that deals with business and environment is the OECD, in which 24 industrialized countries are represented. In addition to these five bodies, there are thousands of organizations which represent common business interests on a sectoral, national, regional and/or worldwide level.

### *The Greening of Business*

There is a growing commitment of business when it comes to environmental issues, especially when it comes to prevention and repair of day-to-day environmental damage on the local and regional level like air, water and soil pollution and prevention of waste. But there is also a growing commitment toward solving longer-term, more global environmental issues like acidification, global warming, depletion of resources, decreasing biodiversity, loss of topsoil and threats to coastal zones.

In the business guidelines from the International Chamber of Commerce and other institutions, some emphasis is given to global issues. This approach certainly has been effective in shaping business attitudes towards these issues. For the next decades the challenge is in finding ways and means to apply these 'guidelines' in formulating strategies and – consequently – in day-to-day decisions at all levels within the business community.

There will be many different ways in which business will respond to global environmental change, in pace, emphasis and attitude. In general, one can expect a change from a re-active attitude – waiting for things to happen – through receptive – still waiting but with an open mind – to a constructive and pro-active attitude. Taking care of the environment becomes part of decision-making on the strategic, the tactical and the operational levels.

The tendency to adapt will probably be strongest in international capital-intensive companies. They are the most resourceful in finding new ways to move forward. Extensive networks of production, trade and services, but also consumers will see to it that the changes in attitude and behaviour of the forerunners will reach as far as the capillaries of the system.

## Earth System Research: Opportunities for Business

The international scientific community has entered into a new and exciting venture; the charting of the systems that support life on earth, the pressures to which they are subjected and the resulting changes of these systems. Business can take advantage of the accummulating knowledge about the Earth System, not only in the scientific results as such but also as a starting point for debate and policy development at both the international and the national levels. The global change research results are important because in the end they will be reflected in either legislation or consumer attitudes.

As already mentioned, we are only at the beginning of grasping the enormous complexity of the Earth System. It is, for instance, a difficult task to separate human influences from the natural variability of the system. Although the models used are complex, they are only crude representations of reality. On top of that, their resolution is low; the outcome as yet is often of limited value for policy making on a national or regional level. Nevertheless, there are compelling reasons for business to establish relations with Earth System research programmes. We list a few of the more important reasons:

(1) *Understanding the sources and flows of information*
If the business community were to be better connected with the primary sources of scientific information, it could better understand the scientific and political debates and even take part in them. A simplified diagram (Figure 1) can help illustrate this point. There are three phases of action: (1) the research itself including its scientific assessments; (2) the evaluation of scientific results in a policy framework; and (3) the action taken. The diagram also shows three types of actors: (1) the science and science-related organizations; (2) the institutions, who digest the results in a policy framework; and (3) national governments (who translate the information into laws, regulations and permit conditions) and business and consumer bodies. It is desirable for business, located in the bottom right compartment of the diagram, to be more directly connected with the sciences in the top left part.

(2) *Shortening the time spans*
Already in 1974 Nobel prize-winners Molina and Rowland published their theory of ozone depletion through CFC emissions. It took almost 20 years to translate their

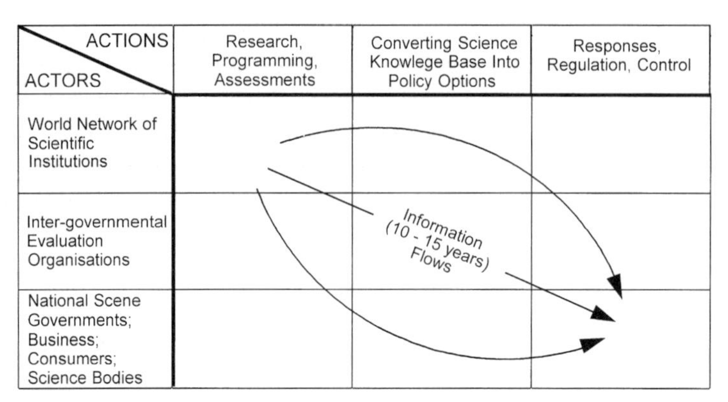


Fig. 1. Schematic diagram showing the sources and flows of information on environmental issues

findings into legislation. The time span between scientific theory and legal action probably will be a lot shorter in the future. Contact between business and science can provide 'early warnings' about new threats to the environment. Acting upon these warnings can save business a lot of money or can create new business opportunities. Like natural ecosystems, the threat to one business is an opportunity for another.

(3) *Preventing distortion*

On its route to the public, scientific information tends to get distorted, partly because it has to be simplified, partly because stakeholders colour the information in a certain way. If business is close to the early stages of production of (scientific) information, there will be more understanding of the how, why and when of the trickling-down process of information and by whom it was altered. The businessman who really wants to know about environmental issues and what they could mean to him, might start by trying to tap into the original information sources.

(4) *Analysing the life cycle of environmental issues*

Environmental issues as well as products, markets and technologies have a life cycle. It often starts with science; the results are placed on the public agenda by intergovernmental bodies and environmental groups. Then the issue is picked up by governments. Advice is asked from interest groups, including business. Then legislation is prepared. Finally, the issue becomes institutionalized as it is more or less brought under control. By getting involved early in the life cycle of an environmental issue, business can influence the agenda setting. That is probably more effective than reacting (defensively) late in the life cycle on issues raised by others.

(5) *Setting priorities*

Identifying the most relevant issues is very important. Companies often not only lack funds and time but, more importantly, they don't have enough people to sort out the avalanche of environmental priority issues coming toward them from government and the public. Admittedly the task of sifting through these issues is difficult. Priorities often change as new knowledge becomes available. Also the attitude of politicians and the public can change on short notice. People who decide the priorities are often criticized by special interest groups, complaining that their views are misunderstood or that they have not been treated fairly. Although a priority list of topics for international environmental research is often a snapshot in time, it can be very revealing. Priorities do change and putting these snapshots in a row creates a film that shows which issues are or will become important. In supporting priority-setting both in research and on the public agenda, business can make a valuable input into the environmental debate.

(6) *Obtaining a helicopter view*
If business is directly connected with the scientific community, it can obtain a helicopter view of a complex area of research. A helicopter sometimes flies high to get a picture of the total area; sometimes it flies low to scrutinize some detail and then rises again a little to search for connections with related issues. A seat in this helicopter can help business to assess the different topics coming towards them through government and public opinion.

## The Science–Business Interface

*The Interface*

Both science and business can benefit from closer relations with each other. The problem is that most of the time scientists do not understand what is going on in business and *vice-versa*. To paraphrase C. P. Snow: 'Between the two [is] a gulf of mutual incomprehension.' To overcome this gulf, it would be sensible to establish fora, both national and international, where science and business can meet each other to discuss matters of global environmental change. The fora could be used, for instance, to discuss priorities for research and to disseminate research results.

The interface between science and business can also be used to search for tailor-made answers to specific business questions. For instance, insurance companies are very interested in changes of climate extremes and the consequences they have on the frequency of storms, droughts, floods and night frosts. Specific results of climate research can help in formulating long-term company policies. (An example of this is to be found in the publication 'Windstorm, new loss dimensions of a natural hazard', the Munich Re-insurance Company.) Without giving definite answers (yet), science can help in providing building blocks for these policies.

*Science-based Consultancies*

The amount of information on global change and sustainable development is overwhelming. There are many national and international research programmes. The enormous output is hardly digestible from a business point of view. Considering the overflow, there is need for a type of consultancy that is strongly related to science, more specifically Earth System Research, including the Human Dimensions of Global Environmental Change Programme. This type of consultancy can provide business with scientific information tailored to its needs. An example is the synthesis monographs published by SCOPE/John Wiley, giving overviews of major environmental issues and opening up roads to new areas of research. (Note, however, that these volumes are not specifically directed towards the business community.) For most business questions, a relatively short study with only a few experts involved would already give many of the answers. Amongst other possibilities, the ICSU network could be used to find the experts wanted in a short time. This type of consultancy could be done by either existing consultancy firms or by new entrepreneurs with a science base and a business outlook (see p. 174). Important for the science community and ICSU is that these types of companies or entrepreneurs should help in creating the necessary science–business interface.

### *The ICSU Connection*

For both the natural and social science communities, it is important to nourish their independence. Their focus is the search for objectivity, maintaining scientific quality and a non-biased assessment of facts. Research has to be safeguarded from outside pressures, while on the other hand, research has to have an open eye for the needs of society.

A good example of this approach is the performance of ICSU, the International Council of Scientific Unions. Through its prestige, it can mobilize global networks of individual scientists. ICSU developed, for example, the international IGBP research framework as described in Chapter 3. This framework is essential in the coordination of national research efforts, avoiding duplication and creating synergies. In this way, scarce resources like money and brains are used optimally.

ICSU is a non-governmental organization with tremendous intellectual strength but only modest financial resources. Its members are the National Academies of Science and international disciplinary Unions. ICSU is mainly financed by its members, so both staff and budget are small. The work is mainly done on a voluntary basis by scientists from many countries. Because ICSU is so successful, the amount of work grows faster than its income. To secure its independence in the future, a trust fund has recently been established in order to facilitate donations from institutions, private persons and business companies. Funds can be allocated to ICSU's core operations or to specific projects (see Appendix 1, p. 172).

## Interviews with Six Senior Members of the Business and Government Communities in the Netherlands

The remainder of this chapter consists of interviews with six senior members of the business and government communities in the Netherlands. The work experiences of the interviewees are varied (e.g., McKinsey & Co., OECD and Shell) and their activities (in some cases, former ones), range from steel production and electricity generation to energy research. Although the interviewees are Dutch, they have had extensive international experience. Apart from their main occupation, they hold, or have held, important positions in the Netherlands, ranging from Dean at a University, through Minister of State, Cabinet Minister and Member of Parliament to President of the Royal Institution of Engineers, with many other involvements in business, education and politics.

These individuals were asked, separately, to elaborate on the following theme: *the importance of international global change research for business, related to their own experiences in the field.*

Although the questions posed were more or less the same, the answers were quite different. Reading the following interviews is therefore a refreshing experience.*

## Interview with Mr. E. Van Lennep

### On financial instruments that can show business and governments the road towards sustainability

'Financial instruments such as an ecotax for $CO_2$ emissions are very useful to get the producer-consumer system on the road towards sustainability – provided that the ecotax meets certain criteria', says Mr Emile Van Lennep, former Secretary General of OECD, and presently Minister of State of the Netherlands. Van Lennep believes in financial stimuli to change behaviour. As a member of the Earth Council and former advisor to the United Nations Conference on Environment and Development (UNCED) in Rio de Janeiro, he is now mainly concerned with the development of financial instruments for a sustainable society.

*Van Lennep*: 'At first, discussions on the subject within UNCED and the Earth Council were rather disappointing. They tended to narrow down to the old confrontation between North and South, where the South is to be funded by the North, this time for abstaining from non-sustainable behaviour. This was disappointing because Rio was meant to bring the discussion on sustainability to a higher level. Another reason why the debate was rather disappointing at first was that people present at Rio thought that they had to find new ways of taxing to collect money for international environmental policies, e.g., a tax on international financial transactions or on postage stamps or a tax on international flights.'

According to Van Lennep, this assumption is wrong for two reasons. The first formal reason is that the creation of taxes is not a matter for the Minister of the Environment, nor for the one responsible for development cooperation. *Van Lennep*: 'The second, more important reason is that through taxing, you would like to discourage behaviour that is harmful to the environment; a sin tax if you wish. I cannot see any reason for a 'sin tax'

* No attempt has been made in this monograph to draw general conclusions from the various ideas expressed by the six individuals. All points of view are valid.

on international financial transactions, or for a tax on international flights. What on earth is the difference in pollution between an inland flight and an international flight, for instance? That is apart from the trouble it would take to enforce these taxes.'

The discussion on these new ways of taxing have distracted attention from the real task; to develop financial instruments that discourage activities that are harmful to the environment. An important instrument is the 'ecotax' on $CO_2$ emissions. *Van Lennep*: 'Such an ecotax is certainly justified, although I am aware of the resistance against it in the business community. It is justified on the grounds that it covers the cost of environmental damage through the price of products and services.'

One of the arguments against an ecotax on $CO_2$ emissions that has been put forward is that the evidence for harmful global environmental change is still too flimsy. There is not yet enough scientific proof of climate change, for example, to justify an ecotax on fossil fuels. *Van Lennep*: 'There is indeed no certainty that the growing concentration of carbon dioxide in the atmosphere will lead to an increase in temperature and a change of climate. But, although there are still doubts, the uncertainty has been reduced lately. Secondly we must realize that science never gives certainties; there is always doubt and rightly so, because if there was not any doubt, you could question the scientific integrity of the research community. The point is that at a certain moment, politicians have to decide if they know enough to take action. In this case, the decision is helped by the fact that we have to go easy on oil and gas anyway, because of finite resources and increasing demand in the developing countries.'

The second argument against ecotaxing of $CO_2$ emissions that has been put forward is that it doesn't help. Some say that people are not driving less even though the price of gasoline has risen 5 cents or more. According to Van Lennep, however, this way of reasoning has been proven wrong; the price of gasoline definitely influences the demand. *Van Lennep*: 'Over the short term, the elasticity is low. People perhaps don't drive less. But over the longer term, elasticity is high. People buy more economic cars or use other means of transport.'

The third argument against an ecotax is that it affects international competition. One country cannot introduce an ecotax without risking damage to its economy. The argument is then extended to a region, Europe, or to the world as a whole. *Van Lennep*: 'Critics of the ecotax have a point there. It would be too altruistic to sacrifice at least part of your national industry for a goal that concerns the world as a whole apart from the fact that it doesn't help because the affected industry will move elsewhere and still produce its share of carbon dioxide. On the other hand, history shows that if you are in the forefront with these type of measures you provoke technical development. Price definitely influences the development of technology. If energy prices rise 30%, a whole range of new technologies becomes available and affordable. In short, countries that are in the forefront with the development of financial instruments are often also in the forefront of technological development and do better in international competition. I have been laughed at for this statement, but I think nowadays more and more people agree with it.'

So taxing can help to get people's noses in the right direction? Van Lennep agrees with that, but wants to make 'more than a footnote' to this statement. 'The ecotax is a tax to influence behaviour, a regulating tax so to speak. Ideally, it doesn't generate any income for government. Governments should not look upon it as yet another source of revenue. The money generated through the ecotax should be ploughed back to stimulate a sustainable economy. The reason for this is that the ecotax is meant to correct a market price that doesn't cover all the socio-economic costs of the product. So it is not a premium for government.'

*Van Lennep*: 'Another remark I want to make is on subsidies. Through subsidies, government lowers prices. Some sectors are subsidized because of social reasons like full employment and income generation, but in a lot of cases the side effect is the introduction of harmful effects on the environment. The Earth Council has started a project to promote policies of abolishing subsidies that are harmful to the environment. You can think of subsidies on energy, agriculture, water and transportation. For instance, as I said before, I am not in favour of a tax on international flights, but I am not against a tax on kerosene. That is the difference between taxing for taxation and taxing for the environment. Through the Chicago Treaty such a kerosene tax is not allowed, so you would have to change the treaty. But in effect not taxing kerosene is a form of subsidy on air transport.'

An argument sometimes raised against an ecotax is that it clouds the PPP (the Polluter Pays Principle). Van Lennep certainly does not agree with that. 'Ecotaxes are meant to reduce harmful effects to the environment. In that sense they are a perfect example of the 'polluter pays principle'. Although PPP dates back to 1972, it is often wrongly interpreted in the sense that it lays the moral responsibility for pollution at the doorstep of the producer. That is wrong in two ways. First of all it has nothing to do with morals. In order to avoid trade distortions, in case some countries subsidize environmental initiatives while others increase prices, a general principle was agreed upon. This principle reflects the most efficient way of paying for environmental damage. Secondly it is not only the producers that have to pay, but also the consumer. It is the producer–consumer system, not just the producer, that harms the environment. Through an ecotax, you can steer the producer–consumer system in the right direction.'

### On Earth-System Research

*Van Lennep*: 'Research into the Earth System is extremely important, because it generates an understanding of our planet. This understanding helps politicians and business in formulating priorities for action. Another thing is that this type of research as far as I can judge helps in creating respect for Earth as an entity in itself, not just as a platform for human life. At the moment I am involved in formulating an "Earth Charter" that establishes the "rights" of the Earth if you could call them that. If these "rights" get a scientific basis, we can make clear the obligations of mankind to the planet.'

'The problem with environmental research is that politicians tend to be carried away with the results of scientific research, as they do also with the results of economic research. I experienced that while I was working as the Secretary General of OECD. Every insider knows that the forecasts of OECD have a certain margin of uncertainty. Politicians therefore often tend to question the usefulness of such forecasts. But they are, to quote a famous Dutch economist 'the least improbable' look into the future. This problem exists, but it cannot be solved, I'm afraid. It is the way things work. Society wants definite answers but science offers sometimes solutions, sometimes doubts. The only thing that a scientist can do about it is to try and uphold scientific integrity.'

### On business responses to long-term global environmental change

*Van Lennep*: 'Although I am not involved in a daily dialogue with industry, I think business is on the right track. The business community has realized in recent years that business interests coincide with the long-term general interest of sustainability. In my opinion there is less resistance against the measures that have to be taken. Business is becoming more and more pro-active; their representatives take part in the debate and that is a healthy development.'

### On other instruments

*Van Lennep*: 'Apart from taxes, there are other financial instruments that you can use to move towards a sustainable society. One of them is "tradeable permits". There are some sectors of economy where you could trade permits on a world scale. You could buy for instance a right to produce a certain amount of carbon dioxide in Holland if you on the other hand pay, let us say, money to conserve the rain forest in Costa Rica. In this way, you could do something other than just scream that the disappearance of the rain forest has to stop. It is an example of the use of environmental issues in international transfers, but, other than a tax on financial transactions, this is a more sensible way of doing things.'

## Interview with Ir. N. G. Ketting

### On the need for agreement on the long-term goals necessary for sustainability

'The concept of sustainable development presupposes a certain unanimity in both the analysis and the long-term solutions of the problems facing us', says Niek Ketting. Ketting is an engineer and president of SEP, the cooperative voice of the electricity-producing companies in the Netherlands. He is also a senator in Parliament and President of the Board of Eulectric, the European association of electricity companies.

*Ketting*: 'The type of unanimity needed is comparable with the motivation with which our Delta flood control works were built. This enormous project was only possible because everyone in Holland agreed on both the analysis of the problem and the solution. Even the people on high ground, 200 kilometres from the shore were determined that the flood of 1953, that killed over 1800 people, was not to happen again.'

But it is said that the problems facing the Earth are an order of magnitude larger than the flooding of a small part of the Netherlands. With the possible change in climate and corresponding sea level rise, even the Delta works might become obsolete. *Ketting*: 'The consequences of the doubling of carbon dioxide concentrations is a very interesting problem, but in my opinion it is not the real problem. The real problem is how to use the energy available as meticulously as possible. By that I don't mean using an economic type of lamp or placing a wind turbine here and there. You have to analyze the whole system; the chain of various functions. Let me give an example. To save energy you can design a high efficiency car that goes 100 kilometres on one litre of gasoline. That type of approach is not enough to solve the problems we are facing. I am not saying that it is useless to design a car like that, but we have to delve deeper. We will have to think through the function of transport in our society – or better; mobility. What is the function of mobility and how can we realize it in the best possible way, taking into account environmental and spatial aspects? Only that sort of functional analysis helps in developing a long-term policy. It shows you which research is still needed, but it also helps in defining priorities and allocating resources. It also confronts you with some very fundamental questions, like "What is mobility and should it be available without limit?" '

'To develop the mobility function within the environmental and spatial constraints requires a consistent policy', Ketting says. 'Consistency means that policies should not change every 4 or 5 years when a new government takes its place. On the other hand, it does not mean that policies have to be rigid. If one element in the system changes, the whole system has to adjust. This type of policy-making requires unanimity in analysis and policy. In short, a vision is needed. The Book of Proverbs says: Without a vision, people sink back into savagery. I think that this also applies to the development of a sustainable society.'

*Ketting*. 'An important part of long-term policy formulation is the inclusion of the behaviour of people'. This subject has hardly been touched upon in research, as has been emphasized in this monograph. 'That is regrettable', Ketting thinks, 'because up till now attempts to make behaviour more environmentally friendly haven't had much effect.'

*Ketting*: 'It is the more regrettable because I believe that the willingness to pay extra for the environment is not inexhaustible. It is a depletable resource like copper ore. You have to be very careful in your appeal to the public, because if the appeals are too many or – worse even – if they are ineffective, you might create environmental scepticism or even cynicism. The only way to prevent that is to make crystal clear why a change in behaviour is necessary. If I as a citizen know that an ecotax on fuel is part of a trajectory towards a sustainable energy supply, I won't like it, but I will regard it positively. If it has all the characteristics of just generating extra revenues for government, my attitude will probably be sceptical or cynical. In my opinion the levy on electricity and gas, to be introduced in 1996 in the Netherlands, would have met with less cynicism if the government had made clear that it was the first step towards a four-fold increase in energy prices to include all costs of using fossil fuel, including depletion of resources and pollution.'

According to Ketting, part of the problem is that leading figures in government and business look at society in a two-dimensional way, the dimensions being economy and technology. That is not enough to explain what is going on, or to try and develop new ways of policy making. *Ketting*: 'Society has more dimensions, a cultural one for instance. Why are people worried about something like biotechnology. There is no economical or technological "rationale" for these worries; instead, they seem to be culturally inspired. Can you explain that and, if so, does that open up possibilities for change?'

Then there is an institutional dimension. *Ketting*: 'Where are these opinions formed? Is it in church? Probably not, but where do people form their beliefs, opinions and attitudes? And in what ways do they affect behaviour? There is still a lot of research to be done in the social sciences.'

*Ketting*: 'A systems approach, based on a vision, also affects the way people and institutions cooperate.' Ketting has experienced that in Eulectric, the European association of electricity companies. 'In Eulectric, there is a lot less discussion on individual member's views on energy saving or reduction of emissions. We don't work with targets that everyone has to meet. The atmosphere is more one of getting the best results with the means available – a shared responsibility to provide Europe with electricity.' In a broader sense, this means that voluntary agreements are more effective than rules, regulations or taxes. Ketting agrees, but to realize this approach, there must be a change in attitude by governments.

*Ketting*: 'We could for instance make a deal with the government to start managing heat in a different way, not using a first-class fuel like natural gas for heating space, but developing "exergy chains", a cascade of applications that starts with a high-valued one like electricity generation and ends with a low-valued one, i.e. heating space. Through these exergy chains, the energy content of gas or oil is used in the most economical way.'

'If we start down a road like that, we cannot predict in advance how much gas we are going to save in 2010 or how much reduction of carbon dioxide we can realize. It is more like teaming up with government, and saying: This is the direction that we think we have to take. We don't know how long we have to walk or how tortuous the road will be, but together we dare to start our journey.'

But don't you expect that for a lot of people this type of thinking and policy-making must be very scary. There are no fixed targets, no regulations to be set – only the conviction that you have to start walking or otherwise it will be too late. *Ketting*: 'The people that are scared to embark on such a journey come up with all kinds of arguments, e.g., that other countries have to do the same. I don't agree. Recent history has shown that a country like Japan, that has a severe environmental policy, does well in international competition.' According to Ketting this type of policy asks for different instruments.

'Regulations are fine, but within a systems approach you have to be able to adjust to changing circumstances. We have such an evaluation system built into our electricity laws. The problem in many instances is that it takes four or five years to change a law. In analysing the system and finding ways of adapting it, you should look beyond national or institutional boundaries.' Ketting's own organization, SEP, has taken initiatives to compensate for carbon dioxide emissions by planting trees, mostly in developing countries. Some people have criticized these schemes. *Ketting*: 'It is difficult for people to accept that you compensate emissions in one area by planting trees elsewhere. In our own country, people feared that we were trying to dodge our environmental obligations. In the countries where the trees were to be planted, there was some resistance as well. Some people thought that we were using this cost-effective way to reach our targets at bargain prices, leaving them the more expensive solutions. And then there were some fears that planting trees would disrupt both the local ecology and way of life.'

These fears and criticism have disappeared. *Ketting*: 'We have shown that planting trees in tropical and subtropical areas is by far the cheapest way of compensating for our carbon dioxide emissions. We knew that all along; it is just a matter of calculations. What we have also shown is that planting trees has positive effects on the communities involved. It has stimulated socio-economic development. On top of that, it has helped in preventing soil erosion.'

'When you think globally, you can spend the money you have for environmental measures more economically. But there are other effects as well', Ketting points out. To illustrate this he describes a project in Rumania. *Ketting*: 'After helping analyse the electricity sector in that country, we are now helping to upgrade one of the power plants, fuelled with lignite. The yield has been boosted from 25% to 35% – not a high level compared with western standards but it means 25% less fuel for the same amount of electricity generated. That means a 25% decrease of lignite in rail transport. That might help to solve other logistic problems like transport of chemicals. So an investment in a power plant has a range of other effects. One of the effects is a stimulus for the economy – a bit more prosperity, which in turn can lead to a bit more social stability. That also leads to sustainability and it might be very important considering the problems that Rumania is facing. To find solutions for these problems is also part of the vision that is needed.'

## Interview with Ir. O. H. A. Van Royen

### Global research contributes to corporate strategy

'Research on global changes as described in this book is an important contribution when trying to develop a corporate strategy', says Ir. O. H. A. Van Royen. 'Maybe not so much the results as such, although they are important, but the insights that arise from it. These insights form the basis of present and future government policies and consumer attitudes, and that is why they are important for companies as well.'

Olivier van Royen is former president of Koninklijke Nederlandse Hoogovens BV and a non-executive director of various companies. He retired in 1993. The name of the company that he worked for and with which he is still associated is Dutch for 'blast furnace'. Hoogovens produces steel and is also a large producer of aluminium. Being one of the largest users of energy, climate research is very important for Hoogovens. *Van Royen*: 'The debate on carbon dioxide leads to changes of energy policy. These changes have a direct influence on our business.'

According to Van Royen, the development of corporate strategy is a matter of putting out feelers to evaluate future threats and opportunities. 'You have to look around, to know what is going on. If you don't, if you only concentrate on your product portfolio, you will lose in the end because developments in society can make or break a company.'

An example of a corporate move based on this type of strategy has been the move into aluminium. *Van Royen*: 'Hoogovens used to make cans of tinplate. Already long ago we saw in the United States that tinplate was pushed aside by aluminium. So we put a lot of effort in tinplate and at the same time we started our own aluminium business.'

A second example of trying to look ahead further than the next meeting of shareholders was the analysis that Hoogovens embarked upon on the future of motor cars, especially the material that future cars were to be made of. *Van Royen*: 'We saw the rise of plastics in cars, so we studied the possibilities of that material. Our conclusion was that an all-plastic car was not feasible – not strong enough. Aluminium seemed a much more likely candidate to partly replace steel in motor cars. That was another reason to step into aluminium production.'

Van Royen does not believe that materials like steel and aluminium will become scarce. 'There is an abundance of ore and bauxite', he says. 'The important thing is that you make as much use as possible of the energy that has gone into these materials. That means recycling, which, by the way, already has been a regular practice in metal production for quite a long time.'

Strategy development is a continuous process. It thus asks for a continuous input of data and insights from all kinds of sources. *Van Royen*: 'Making steel and aluminium for instance is changing energy into metal – lots of energy. When oil and gas become scarce, you want to produce aluminium on a site where there is an abundance of cheap energy like hydropower, flare gas or on top of a coal mine. So you can imagine that Western Europe in the long run is not the ideal place for producing these materials. That is an important conclusion for a company like ours. It does not mean that you are going to move tomorrow. A blast furnace is not something that you move as easily as a factory that produces French fries. But setting out a course towards the future and evaluating threats and opportunities, you have to keep in mind long-term development in the supply of energy. Especially for a very capital-intense industry like ours, it is very important to look far ahead to see what is coming towards you, such as global environmental change'.

'Predicting the future is quite difficult', admits Van Royen. 'We did not see the changes coming in Eastern Europe', he says, 'nor the possibility that these changes would lead to dumping of aluminium on our markets. But that does not mean that you don't have to keep on trying to get a view of what is in the crystal ball.'

When asked, Van Royen says that developing a strategy is, for a large part, intuition. 'Scientific and technological research plays a role of course. It is quite important even in reducing uncertainties. But the bottom line is that you have to base your strategic decisions on your own evaluation of facts and developments. Calculations help, but don't give the complete answer. Take for instance the greenhouse effect. You want to know how scientifically sound these predictions of a rising temperature are. I must say that I am not too sure about that. There are still a lot of complicating factors like the amount of water vapour in the atmosphere or the role of the oceans. Then there are the political effects. In what way will government policy change due to the greenhouse effect, whether it is real or not. You have to take all these things into account. That means you rely on educated guesses, not on science; that is difficult. We have still not succeeded in predicting the value of the dollar a week or a month ahead. So in developing a long-term strategy, you have to guide yourself by a healthy scepticism.'

One of the societal responses of a possible global change in climate is the introduction of a tax on energy. Van Royen doesn't think much of that. 'A lot of studies have shown that the price elasticity of energy is too low for an energy tax to have much effect. You have to raise the price by 20% to get people to reduce energy consumption by 1.5%.' Van Royen thinks a tax on energy is peanuts if there really is a greenhouse effect. If it is true that higher atmospheric concentrations of carbon dioxide lead to climatic change and sea level rise, then you have to intervene on a massive scale: strict measures to get people and companies to save energy; re-introduction of nuclear power; and fast development of renewables like solar, wind and biomass. It is either that or sit and wait, which is in my opinion not a very good government policy. On the other hand, if there is no greenhouse effect or sea level rise, you don't have to introduce a tax on energy as is proposed in the Netherlands for 1996. It is quite a waste of money, I think. Apart from that, the preoccupation with carbon dioxide distracts attention from other important issues.'

Van Royen, who is also a member of the Dutch Advisory Council for Energy, is of the opinion that a tax on energy is not the way to decrease demand. 'The real solution', he says 'is innovative technology. Two hundred years ago Malthus predicted starvation on a global scale, because the Earth's resources would not keep up with population growth. The Industrial Revolution based on new technology proved Malthus wrong. I think that will happen again. New technology opens up the possibility of overcoming the problems that we now face. Whether it is nuclear fusion or something else is hard to tell. The only thing you can say is that you have to invest in research and development. It will pay off, although you don't know in what way yet. That is why it is such a pity that governments are cutting back on expenditures for research. It also cuts off a lot of possible solutions.'

'If you really want to put a tax on energy, I would not mind on the condition that the revenues are used for research and development of new sources of energy. A tax of 1.5 or 2% of the energy price would create a nice stimulus. This would be better than just using the revenues to decrease income taxes or something like that as is now considered in the Netherlands.'

Van Royen puts a lot of trust in technology development. 'Considering past experiences, that is certainly justified', he says. 'You know what the problem is: there are so few engineers in politics. Not in the sense that you can run society like a machine, but because politicians lack insight into the possibilities of technology.'

## Interview with Ir. J. M. H. Van Engelshoven

### Research can help avoid conflicts of interests

'Biomass can be an interesting source of energy and raw material for the chemical industry', says Van Engelshoven. 'But you have to avoid a conflict of interest with food production and conservation of biodiversity. All three goals put a claim on land. In a variety of ways, research can help in optimizing these claims.' Ir. Huub van Engelshoven is a former managing director of the Royal Dutch Shell Group and now president of the Royal Institution of Engineers. 'At one time in my career, I was responsible for the planting of twenty million trees per year. Sites were selected on three criteria; they should be unfit for food production; they should be largely uninhabited: and they should not contain valuable ecosystems like rain forests.'

### Reforestation in developing countries

One of the criteria mentioned above almost got Van Engelshoven into trouble. 'The Prime Minister of Malaysia told me not to interfere with the logging policy of his country, especially not since we Europeans have deforested our land to build ships to colonize their land. Still we stuck to our policy and in the end the Prime Minister and I got along very well.'

Even when applying these criteria, there is a lot of land available for the production of wood. Part of the wood can be used for energy production or for the production of paper. Another important application is wood for building. The problem is that a fast-growing tree like the eucalyptus is too soft to be used as wood for building. *Van Engelshoven*: 'In the Shell laboratory in Amsterdam, we have developed a process to harden softwood. It is amazing. When you apply it to poplar wood, for example, it feels like mahogany. In my opinion, this preservation technology is a good example of how science can help in preventing a conflict of interests; in this case, a conflict between land claims for wood for building and biodiversity. On top of that, it helps save rain forests. The technology is now commercialized by a third party.'

### Energy supply

Van Engelshoven's concern for global change is rooted in his career with Shell. That is why he is concentrating on the possible changes in energy supply and demand and the societal effects they have. 'Looking at energy supply, I distinguish between sources of energy and systems to make these sources available to an ever-growing world population in a way that does not affect the Earth System. In my view, there is no shortage of energy sources. There is still plenty of coal in the world. No scenario I know of predicts a shortage of oil and gas within the next 50 years. Biomass as a source of energy is only at its beginning; the yields per hectare can be boosted if we put some effort in it. The same goes for geothermal energy. Solar energy is almost unlimited and we hardly have begun to tap sources like wind and wave power. Nuclear energy is a political choice, with different positions taken between Germany and France, for instance, or Japan and the United States. But the energy is there when you need it and it is probably a lot safer than coal as a primary source of energy. I do not yet include nuclear fusion in my list, because I fear it will always stay a promising technology. Anyway, adding it all up we have quite a few energy sources. The world may go under, but certainly not from a shortage of energy.'

Nevertheless, a lot of people fear shortages. *Van Engelshoven*: 'The problem is not the lack of energy sources but how to make them available. One important consideration is that the price of fossil fuels, oil, gas and coal, will probably stay at the same level. Since 1900 these prices have been more or less constant if you take the general price increase into account.'

### Free choice

That would mean that we can still rely on fossil fuels for a long period? *Van Engelshoven*: 'We could, but that would not be very sensible. There are two reasons why we have to switch from the "easy" fossil fuels to other energy sources, preferably renewable ones. The first reason is that sustainability also means that you give future generations the same freedom of choice we have enjoyed. If we burn fossil fuels for another 100 years or so, the choices for the generations living then will be very restricted. The second reason is the environment. Apart from the possibility of a change in climate caused by carbon dioxide – of which I am not so sure yet – there are the emissions of nitrogen oxides and sulphur dioxide that we have to take into account. So I agree with people who say we have to cut down on consumption of fossil fuels.'

You said that you are not so sure about climate change? *Van Engelshoven*: 'Let us say that I do not agree with people who say that we have to switch energy sources because of possible climate change. If we are indeed threatened by a change in climate due to emissions of carbon dioxide, we still have 30 years to plant trees in the Sahara desert to fix the surplus of carbon dioxide.'

### Making renewables competitive

How can you make renewables competitive? *Van Engelshoven*: 'There are two questions involved. One is how to make renewables competitive and the other one is how to adapt them to local circumstances. The questions are related. I would imagine that introducing solar energy is economically more feasible in Burkina Fasso than in a country like the Netherlands, which already has an extensive infrastructure for transportation of electricity. The answer to both kinds of question is: a lot of research and development. For instance, I could imagine that you develop a roofing for houses in the Netherlands, made of solar cells that produce electricity and that have a payback-period of six years. Not only that, but to make optimal use of the solar electricity that you capture, you also design a system to link these roofs with the national grid. There are a lot of "if's" in that and to solve them, you need research, both fundamental and applied research and technology.'

### We should all be ashamed!

But governments are cutting back on research expenditure. *Van Engelshoven*: 'Indeed they are, and I think that they are taking the wrong route. They are actually not unique in that respect. Both government and industry sometimes have a short horizon. If you want to develop a certain technology, it needs a period of nursing which might well last more than ten years. Viewing the problems that society faces nowadays, we are going to need a lot of research and development – and not only for developing new ways of converting energy. Another important area for research is energy conservation. I am not exaggerating if I state that we can save up to 75% of the energy we now use, without giving up anything in terms of comfort and quality of life. We should actually be ashamed of ourselves, wasting so much energy. Part of these savings can be realized by readily available technology like energy-efficient cars and light bulbs and refrigerators that use only half of the electricity that they use now. Part of the savings needs a problem-solving type of research. If government and business invest in technology that helps save 2% per year, we will have saved 70% by the year 2030!'

## Interview with Prof. Dr. Ir. H. H. Van Den Kroonenberg

### Designing energy systems for a sustainable future

'We are engineers, so we translate global systems research into new technology', says Harry Van den Kroonenberg, 'not just nuts and bolts, but we try to design future systems for a sustainable society.' Van den Kroonenberg is the director of the Energy Research Institute ECN in Petten. He is a former dean of Twente University and president of the Netherlands Society for Industry and Trade, founded in 1777.

A few years ago Van den Kroonenberg's institute, ECN, embarked on an ambitious programme, called ENGINE (Energy Generation in the Natural Environment). The aim of the programme is to design an energy system that is both clean and safe and on top of that, sustainable. *Van den Kroonenberg*: 'Sustainability means the use of renewables like solar, wind and biomass and a potentially clean source like nuclear energy. We have to realize, though, that for the next fifty years or so, fossil fuels will still play an important role. Our energy supply will surely consist of a mix of renewables, fossil fuel and nuclear. The goal of the ENGINE programme is to design an energy supply mix that is both sufficient and does not lead to harmful emissions to air and water or to unmanageable waste.'

Hydrogen and electricity play a key role in the ENGINE-approach as a clean, secondary source of energy. *Van den Kroonenberg*: 'You can of course use natural gas for heating a single house or burning petrol for moving a single car. Preventing pollution though is much more complicated and less cost-effective if you have a lot of small, decentralized units than if you have a few large units. The same goes for nuclear fission. If you use hydrogen and electricity as a secondary source for either transport or heating, the pollution from decentralized use is almost zero. The only thing you get when you burn hydrogen is water.'

So development of energy systems certainly has to be done on an international level? *Van den Kroonenberg*: 'Yes. We started within our own institute but we are now in the process of internationalizing it. It is no use limiting yourself to the local situation. The mix of energy supply systems will certainly differ from country to country, but the research and development must be done on a global scale. Inventing the wheel over and over again is a waste of a very precious resource, namely brain capacity.'

'Another reason for a global approach is that what you do in one region affects developments in others. The effort we put into reducing emissions of carbon dioxide might be useless if countries like India and China are really going to use their enormous supplies of coal. Whether or not these countries use their coal depends on the type of choices they make. India for instance, has not only large reserves of coal, but also a great potential for using wind and solar energy. It might be more interesting for them to tap these resources, at least in certain areas, than using fossil fuels. The advantage of using these sources is that you don't have to build up an infrastructure for electricity distribution for the whole country. In areas where there is no grid, it is now already more economical to convert sunlight directly into electricity than extending your national grid.'

This type of decentralized energy supply has other advantages as well. It is a quick way of bringing electricity to even the most remote areas. *Van den Kroonenberg*: 'What is happening in Indonesia, for instance, is that a small photovoltaic system lengthens the day for people. The darkness of the evening is gone, so people can accomplish more; children can do their homework. In a way, these solar cells also create a link with the outside world through radio and television. That can be a cultural enrichment. Then there is the possibility of cooling. Once when one of our employees was in Africa, he came across a little hospital in the forest. The doctor used to keep his medicine in one of these old fridges that run on paraffin. If the shop ran out of the stuff, the medicines were not cooled for a while and people actually died. We solved the problem with one square meter of photovoltaic cell and an electric fridge.'

Development of energy systems is not limited to place or time. The things you do now, affect not only your neighbours, but also your grandchildren, i.e. the energy systems of tomorrow. Van den Kroonenberg illustrates this by pointing to a project he came across in Thailand. In the north of that country, refugees from Cambodia were cutting trees for cooking. To prevent the devastation of the forests, the Thai government started a project whereby large amounts of waste chaff from rice were mixed with oil and then shaped in the form of tree trunks. *Van den Kroonenberg*: 'This solution was not only important in solving the immediate cooking problems of the Cambodian refugees without destroying the forest, but it also showed the way for future developments, more specifically the use of biomass for cooking. It fits into the type of systems approach that we want to develop.'

A systems approach further shows that solutions are not good or bad *per se*. It all depends on how you make them and use them. You have to look at the whole chain, from cradle to grave. *Van den Kroonenberg*: 'Take for instance renewables like wind and solar power. A wind turbine or a photovoltaic installation does not produce sulphur dioxide or radio-activity. Still, neither is *a priori* a clean source of energy. The production of photovoltaic cells causes a lot of waste. And I am wondering what we are going to do with all these thousands of turbine blades. They are now mostly made of fibre-reinforced polyester or other type of plastic. From a sustainability point of view, it might be better to use wood or aluminium, a material that can be recycled easily.'

Up till now, Van den Kroonenberg has not mentioned nuclear fusion as a possible part of an energy supply system. The reason is that he does not believe that production of electricity through nuclear fusion is feasible within the next fifty years. 'If we want a sustainable society in 2030 or thereabouts, nuclear fusion will be too late', he says. 'Since the fifties, the point in time at which fusion would be available as a source of energy has been postponed every five or ten years. I am afraid it will be postponed some more. Apart from that, present fusion is not a clean technology by itself. The materials of the reactor are bombarded by neutrons and become radioactive themselves. They have to be replaced every few years, so you have quite a heap of radioactive material you have to get rid of in a decent way. There is an alternative; a fusion reaction between helium-3 and deuterium. That fusion process produces energy without the neutrons, so you might call that clean. As far as I know, there is not a lot of research being undertaken on this type of fusion.'

The ENGINE programme, with a yearly budget of around 10 million Dutch guilders, is based on an approach called 'back-casting'. *Van den Kroonenberg*: 'Instead of taking the present day situation as a starting point for forecasting, you imagine yourself being transported with a time machine to the year 2030. You try to guess what type of technology is needed in that year, knowing that there are about 10 billion people in the world that need at least food, clothing, shelter, health care and safety and preferably a bit more. This is to be achieved under the condition that the carrying capacity of planet Earth is not exceeded. That is why the type of research described in this book is so important. You then look back to 1996 and try to figure out what type of technology must be developed to meet the needs of 2030.' See Appendix 4, pp. 194–6, for a discussion of scenario building.

'In presenting these technologies, you often get criticism', Van den Kroonenberg says. 'People say that although it is technically feasible, the yield is too low. Photovoltaic energy still is confronted with that type of reasoning. Other people will point out to you that the economics don't work out well. Okay, so be it. My answer to this criticism is that it is our job to make it technically and economically feasible, at least to do our utmost.'

This type of technology development, that is based on a view from the future, can only be done when people who fund the work are convinced of the importance of it. According to Van den Kroonenberg, that poses a problem nowadays. Governments tend to be short-sighted; their goal is to survive after the next election. And business is retreating to its 'core'. *Van den Kroonenberg*: 'Even a rich country like the Netherlands is cutting down on expenses for research and development. Last year we had to cut back heavily on our research both in nuclear and renewables. The sad thing is that cutbacks on a national level also have consequences for the international programmes in which we participate. It is a sort of multiplier effect which more than doubled our loss of funds.'

Van den Kroonenberg is not a man who keeps lamenting about loss of funds. He is already trying to find other ways to generate money for the institute. One of them is to combine long-term research, like the ENGINE programme, with trouble-shooting. *Van den Kroonenberg*: 'The spin-off of our research into energy systems might be very useful for medium and small-sized companies or other customers. For instance, our research on the use of biomass has produced results that will now be used for a 20 megawatt power station fuelled by gasification of wood. That is the nice thing about this kind of systems approach. You can develop now at least part of the technologies you will need tomorrow.'

## Interview with Dr. P. Winsemius

### Analysing the obstacles to change

'When you listen to people, they always make the right sounds about the changes necessary for sustainable development', says Dr Pieter Winsemius. 'Everyone agrees that we cannot go on polluting the environment and destroying ecosystems. The interesting thing is, although we are all aware of the necessity for change, nothing – or at least not enough – is actually changing.' Dr Pieter Winsemius is Director of McKinsey & Co. He is also President of the Vereniging Behoud Natuurmonumenten, the largest association for nature conservation in the Netherlands. From 1982 until 1986 he was a Cabinet Minister for Housing, Physical Planning and the Environment.

Winsemius likes to compare the reactions to change with a journey through different landscapes. 'The promoter of change, the one that preaches the need for change towards a more sustainable society, first has to travel through the "Desert of Good Will". People fully agree with his analysis and long-term solution, but they don't do anything about it. Then, when ideas are – partly – converted to regulations or changed attitudes, he has to pass through the "Mountains of Anger". People will say he is mad, proposing this or that. After that, he usually enters the sunny "Valley of Righteousness". There live the people who always said that this was the right thing to do. By saying that, they actually discredit the promoter of change, who, as the prophet, has been crying in the desert. But there is no need to feel aggrieved by that; it just happens.'

### Barriers

According to Winsemius, there are, in general, five barriers for change. 'First of all', he says, 'there is the short time horizon of many people – not only in politics, but also in business. Most of the attention goes to pressing issues; these issues are urgent because of either their physical or temporal proximity. You see this happen on the shop floor, but also in the board room. When it comes to environmental issues, senior managers often choose a strategy of staying out of trouble. That means complying with regulations and avoiding incidents. Because of that, they remain preoccupied with operational issues, like waste disposal and emissions to air and water. Environment is looked upon as something for technicians; it belongs in the same category as health and safety. Because of that, most societal leaders, whether in government or in business lack a long-term perspective on the major environmental issues. In fact they tend to fall victim to "issue overload".'

### Climbing

A second barrier to change is the inclination to overestimate both what has already been accomplished, and what has to be done yet, to satisfy those 'environmentalists'. *Winsemius*: 'You can compare it with someone who is at the halfway point on a climb up a mountain. Looking down he is impressed by the distance he has already climbed. He is also fully aware of the effort it has taken to get there, an effort, by the way, that is never duly appreciated by others.'

'Looking up, he realizes there is more to come. The "issue overload" mentioned above, makes matters even worse so that the climber loses sight of the top. Similarly in business, management tends to overestimate environmental expenditures by a factor of two. Halfway up the mountain, the overestimate is at its highest, creating a psychological barrier to do more than necessary.'

### Lack of understanding

A third barrier to change is a mutual lack of understanding between business on the one side and (parts of) government, science and NGOs on the other. *Winsemius*: 'In general, companies understand too little about the motivations of external players and vice-versa. You don't necessarily have to agree with each other, but you have to at least understand each other's position. If you don't, it becomes all too easy for parties to resort to adversarial positions, which in turn, prevent an integrated response to the challenges awaiting us.'

### Time lags

Internally, companies are confronted with lagging structures and systems, the fourth barrier for change. *Winsemius*: 'Systems and structures of companies are adapted to yesterday's needs, creating institutional convervatism. That is all right when change is slow, but when the situation is changing rapidly, company responses are often too slow.' As an example of this lag, Winsemius points out that line managers are not rewarded for superior environmental performance: on the contrary, sometimes. Also, the environmental information systems are compliance-oriented. The constant stream of data on emissions and listings of inspections are largely irrelevant for measuring progress or developing long-term strategies. *Winsemius*: 'For that, you need insight, not just data. The problem is that a lot of companies haven't developed the external antennae necessary to get the information required to gain these insights.'

### Site managers with pivotal responsibilities

The fifth barrier to change is the apprehension of people holding pivotal jobs, i.e., positions where the vision from the top is translated into operational action. In most cases these are 'site managers'. These people hold the key to change but they often lack either the skill or the will to execute a major change. *Winsemius*: 'They are, and sometimes rightly so, quite sceptical about the board's commitment to environmental goals – a scepticism that grows out of the criticism they have to endure from both the shop floor and the outside world.' To illustrate this point, Winsemius shows a graph of the results of a questionnaire by McKinsey, in which managers were asked whether they agreed or disagreed with the following statement:

> We are successful at getting production managers to feel comfortable with safety and environment.

Of top management, 100% agreed with the statement. Of site managers, only 9% agreed. *Winsemius*: 'Communication between top and middle management is often garbled, leading to insecurity among middle managers. When people are insecure, they tend to do nothing out of line.'

### Behaviour

How can science help to overcome these barriers? *Winsemius*: 'Science produces a lot of information. But to use this information in working towards a sustainable society, you need the social sciences. People have to change their attitudes and behaviour. As a physicist by education, I consulted some social scientists I know. One of them, the Dutch sociologist Kees Schuyt, told me about the several ways there are to get people to change their behaviour. One way is by laws and regulations. The effects are visible within one to five years, at least if you put some effort into enforcement. But regulation is very tiresome. On top of that, it might create an atmosphere that you don't want, instilling fear in people.'

'The second way is to try to get institutions to change. That type of change is much slower; it will take you anywhere from 15 to 30 years. Twenty years after the UN Conference on the Environment in Stockholm, you see a shift in industry's attitude towards the environment. Instead of being told what to do by laws and regulations, companies now take initiatives because they feel that that is the way they should behave. You no longer throw away your waste in the woodlands, not because it is forbidden, but because you don't want to hit the front page as a polluter.'

'The third way to get people to change is through a change of values, norms and standards. People take responsibility for their own behaviour. It is a process that takes between 30 and 100 years. But it happens. For instance, I have no ash tray in my room, not only because I don't need one, but also because smoking is becoming an indecent habit. And if I had proposed ten years ago as Minister of Environment, that citizens should separate out household waste, I would have been hanged. Now it is normal practice in Holland. Societies do change, although it may take a while.'

### Maslow's hierarchy

*Winsemius*: 'To accelerate the process of change, we first have to understand what is going on. An important tool of analysis is the hierarchy of needs as formulated by the American psychologist Abraham Maslow, who identified five levels or stepping stones that every individual follows in fulfilling his needs. The first one is physiological – food, shelter and so on. The second level is safety and security. On the third level we find acceptance by others. The fourth level is self-respect and the fifth and highest level is self-actualization. Every stepping stone can only be reached after the needs at the previous stepping stone are met. You can translate that idea to society. If in a given society too many people are still trying to fulfil their needs at a low level (e.g. the basic need for food or shelter), imperatives of that society are dominated by that unfulfilled need.'

### Environmental 'needs'

Winsemius relates the Maslow hierarchy of needs to the development of environmental 'needs' of society. 'On the first level, people try to fulfil their basic needs; there is no environmental policy to speak of other than, for instance, rodent control.'

'On the second level of safety and security, attention is given to threats to human health. Environmental policy consists of supplying clean drinking water to prevent disease; better housing; construction of sewage systems; and so on.'

'The third stepping stone is acceptance by others; the need to belong to a community. When societies rise to this level, concern shifts to the well-being of the community, which is threatened by air and water pollution and the destruction of nature. Maslow's need of "belonging" is expressed in a greater concern for the quality of shared surroundings.'

'On the fourth stepping stone [(self-)respect], people stop thinking solely in terms of threats to their health and surroundings. Environmental quality, including ecological balance, becomes an objective in its own right. Although it is still not easy to be green, people are willing to pay extra for green products and services, thus creating opportunities for the "greening" of business.'

The fifth and highest level is the level where all players recognize the necessity to view ecological security as equal to socio-economic security and resource security. According to Winsemius, this stepping stone can only be reached when the basic needs of the world population as a whole are largely met. *Winsemius*: 'If the needs belonging to the previous levels are not met, human behaviour arising from the pursuit of these more urgent needs will still result in unsustainable pressure in the form of over-fertilization, deforestation and exhaustion of resources. To go from one stepping stone to the next, you have to overcome the barriers for change that I mentioned earlier. These actually seem to get bigger, the higher you climb up Maslow's steps. That probably also explains why the work on Global Change research hasn't caught on yet in industry. The research, by its nature, is concerned with problems that arise from the needs on the fifth Maslow level. Industry, at least in the OECD countries, is still reacting to needs arising from the third and in some cases the fourth level.'

### Sounding board

'That doesn't mean though that the research on Global Change is useless for industry. Far from it', says Winsemius. 'The problem is that industry does not see the use of it yet, blinded as they are by today's problems.' According to Winsemius, a sensible thing for companies to do would be to install an environmental advisory board consisting of heavyweights in the environmental field. Once or twice a year, you ask their advice on certain matters concerning long-term strategy of the company, e.g., on the depletion of raw materials used by the company and the possible alternatives. This environmental advisory board should not concern itself with day-to-day problems or technical details. It would function instead as a green sounding board for top management.'

## Selected References

Business Council of Australia (1994) *Meeting the Environmental Challenge*, 57 pp.

Cogan, O. (1992) *The Greenhouse Gambit: Business and Investment Responses to Climate Change*, Investor Responsibility Research Center, Washington, DC, 484 pp.

Ditz, D., Ranganathan, J. and Banks. R. D. (eds) (1995) *Green Ledgers*. World Resource Institute, Washington, DC, 181 pp.

Frause, B. and Colehour, J. A. (1994) *The Environmental Marketing Imperative: Strategies for Transforming Environmental Commitment into a Comparative Advantage*, Probus Pub., Chicago, 264 pp.

Schmidheiny, S. (ed.) (1992) *Changing Course*. MIT Press, Boston, MA, 374 pp.

Socolow, R., Andrews, C., Berkhout, F. and Thomas, V. (eds.) (1994) *Industrial Ecology and Global Change*, Cambridge Univ. Press, 500 pp.

UNEP (1996) Production and consumption, *Our Planet*, **7**, No. 6, 1–36.

# 9. Environmental Non-governmental Organizations (ENGOs)

J. W. M. LA RIVIÈRE[1], R. E. MUNN[2] and P. TIMMERMAN[2]
[1]*IHE Delft, The Netherlands;* [2]*Institute for Environmental Studies, University of Toronto, Canada*

## Introduction

Over the past 25 years, increasing numbers of public interest groups concerned with the environment have emerged locally, nationally, regionally and internationally. At the 1992 Rio Conference, over 1400 non-governmental organizations (NGOs) were officially accredited and represented. These had mandates in the area of the environment, development, or both. Their influence on public opinion and on the political will of decision-makers is felt by many to be considerable. Quite often of course, NGOs represent differing points of view, e.g., those wishing to stress poverty and development concerns vs. those wishing to protect the environment. It should also be noted that since Rio, there has been a substantial increase in the number and power of Southern NGOs.

**Two Early Examples of Environmentalism**

(1) During the XVth century, the task of locating Eden and re-evaluating nature had already begun to be served by the appropriation of the newly discovered tropical islands as Paradises. Dante's siting of an 'Earthly Paradise' in a 'southern ocean' is a case in point.

(2) By the early 1860s, anxieties in Britain about artificially induced climatic change and species extinctions had reached a climax... . The subsequent evolution of the awareness of a global environmental threat has, to date, consisted almost entirely in a reiteration of a set of ideas that had reached full maturity over a century ago.

Richard Grove, *Nature*, **345**, 11–14 (1990).

Of course, the roots of environmentalism go back many centuries (Grove, 1990), and Charles Darwin, Henry David Thoreau and President Teddy Roosevelt provided inspiration for environmentalists in their times (see the Box above for two other examples). More recently, Rachel Carson in the United States and Barbara Ward in the UK were influential in the environmental revival of the 1960s, which led to the United Nations Stockholm Conference on the Human Environment in 1972.

*R. E. Munn, J. W. M. la Rivière and N. van Lookeren Campagne (eds), Policy Making in an Era of Global Environmental Change, 165–168.* 

## Legitimacy of Environmental NGOs; Their Place in Society

The role of most Environmental NGOs ((ENGOs) is that of a lobby or pressure group that uses available public information channels (TV, radio, newspapers, pamphlets, etc.), public demonstrations, and discussions with policy-makers and business leaders to present their cases. Because they often have substantial expertise, particularly in OECD countries, the more professional ENGOs are currently being used as unofficial consultants to regulatory bodies, assessment panels and government ministries.

In a number of countries in the developing world, ENGOs often represent the only serious opposition to authoritarian regimes. A familiar example of this in the 1980s was the use of environmental groups in Eastern Europe as rallying points for anti-regime activists.

The question of how (or whether) ENGOs can claim to speak on behalf of the 'public interest' is complicated. The officers of most ENGOs have not been chosen through democratic elections, and are only accountable to the membership through the payment (or withholding) of annual dues, or by the occasional use of petitions, public meetings and other forms of keeping in touch with the members. Ironically, most ENGOs insist that government bodies be democratic and transparent. Environmentalists consider that ENGOs are acting in the larger public interest, and resist being considered as 'special interest' groups, as they see the long-term sustainability of the planet as being in everyone's interest. Similarly, lobby groups for business or other interests argue that special consideration of their concerns will bring improvements in general social and economic well-being.

One of the roles that ENGOs have had to play on the international scene is as a voice for the citizen concerned about the global environment in spite of the local interests of one's own nation-state. The ENGOs argue, for example, that some national representatives in the United Nations do not reflect the views of their citizenry. There is, of course, no provision for democratic processes in the United Nations, and this makes it very difficult for public pressure to be brought to bear on the decisions of UN bodies, except indirectly. The Rio Summit and other recent conferences have confirmed the wisdom of involving ENGOs more directly in the process.

The relationship between environmental groups and political parties is different in different countries: in some, ENGOs have fostered close relations with major parties, e.g., in the Netherlands; in others, ENGOs have (with limited success) allied themselves with 'Green Parties', e.g., Germany, while in still others they have stayed away from such alliances, e.g., in Canada. At the same time, Ministries of the Environment have been created in many countries, but this has not always given the environment the all-pervasive position in government policy formulation that it deserves. In this regard, the ENGOs have helped focus public attention on environmental issues.

Of course, ENGOs require good documentation, which means that they are important 'clients' of environmental research.

## Environmental Advocacy

Most scientists are cautious, and more than a little afraid of the media who tend to highlight the most speculative part of a press briefing as being most newsworthy. However, a few scientists believe that their role is that of advocate. Sometimes, scientific advocates are beneficial, alerting the public to environmental threats long before the scientific evidence is persuasive, and resulting in increased research funds being made available to universities and government laboratories. Issues that were influenced in this way are acidic deposition in Sweden in the 1960s (Svante Odén alerted the Swedish people to the fact that many rivers and lakes had become more acidic over the previous decade), and stratospheric ozone depletion in the 1970s (the media in the United States picked up the connection between ozone depletion and CFCs in aerosol spray cans).

There are dangers in environmental advocacy, however. The credibility of the scientific community suffers when environmental advocates express extreme views. Particularly when the threats are long-term with uncertain consequences, the danger of exaggeration becomes very real. This is part of a more general problem of communication. According to Rowland (1993), the public lack of confidence in the environmental information provided by scientists, business and government is due to faulty communication. 'Each of us is bombarded by messages from television, radio, magazines, newspapers and so on, but the science contained therein is often badly garbled' (Rowland, 1993). The solution according to Rowland (1993) is through public education, 'and at a lower level than we would like to see'. One goal of the ENGOs, in partnership with governments, business and scientific bodies, should be to foster a scientifically literate society.

## Relationships Between Business and ENGOs

There are various ways in which ENGOs and business can engage in productive partnerships. To begin with, ENGOs are important sources of local information about

**A Lack of Scientific Consensus – or the Propagation of Misinformation?**

Mario Molina, who has been in the forefront of stratospheric ozone research since the early 1970s, was invited to give a lecture on ozone depletion at the Rio Conference in June 1992. Much to his surprise, the previous speaker argued that CFCs were not the cause of ozone destruction in the stratosphere. So much chlorine was getting into the atmosphere from sea spray, volcanoes and biomass burning, according to that speaker, that CFCs could play only a minor role. Molina was stunned, and remarked later that 'he was not going to be able to teach the audience in half an hour'.

Gary Taubes, *Science*, **260**, 1580–1583 (1993).

public worries, and they often provide 'early warnings' of issues that will be of concern to corporations in the near future. It was environmentalists who predicted, and provoked, the movement towards the recycling of materials, especially newspapers, in many industrialized countries. Those companies that ignored the environmental movement and failed to modernize have had great difficulties.

Because ENGOs have more freedom to manoeuvre than some companies and governments, they have often been vehicles for experiments in the fields of energy, housing and other prototypes that cannot obtain conventional support. It is also the case that ENGOs have supported companies that promote environmentally friendly actions in their efforts to put pressure on governments and other companies. These informal alliances are found in many industries, and also play a part in international negotiations over industrial standards for international trade. In a few cases, ENGOs and business have formed alliances, free of government bureacracy, to resolve difficult issues such as regulation of toxic wastes.

A number of ENGOs, such as the World Wildlife Fund, have a long tradition of corporate sponsorships, which includes developing mutually beneficial partnerships. Examples of these include partnerships to save animals associated with company logos – the Canada Life pelican; Jaguar's jaguar – as well as various oil company campaigns to enhance sea life around drilling platforms.

## Selected References

Bramwell, A. (1989) *Ecology in the 20th Century*. Yale University Press, New Haven, Conn.

Choucri, N. (ed.) (1993) *Global Accord: Environmental Challenges and International Responses*. MIT Press, Cambridge, MA.

Grove, R. (1990) The origins of environmentalism. *Nature*, **345**, 11–14.

Jamieson, A., Eyerman, R. and Cramer, J. (1990) *The Making of the New Environmental Consciousness: A Comparative Study of the Environmental Movements in Sweden, Denmark and the Netherlands*. University Press, Edinburgh, Scotland.

Lipschutz, R. D. and Conca, K. (eds) (1993) *The State and Social Power in Global Environmental Politics*. Columbia University Press, New York.

McNeely, J. A. (1993) From science to action: what is the role of non-governmental organizations? In *Proc. Norway/UNEP Expert Conference on Biodiversity*. Trondheim, Norway, 24–28 May, 1993, pp. 162–165.

Paehlke, R. C. (1989) *Environmentalism and the Future of Progressive Politics*. Yale University Press, New Haven, Conn.

Rowland, F. S. (1993) President's lecture: the need for scientific communication with the public. *Science*, **260**, 1571–1576.

Taubes, G. (1993) The ozone backlash. *Science*, **260**, 1580–1583.

Young, O. R. (1994) *International Governance: Protecting the Environment in a Stateless Society*. Cornell University Press, Ithaca, NY, 221 pp.

# Appendix 1. Contact Addresses and Further Information on Authors, Supporting and Sponsoring Organizations

## Addresses of Contributors to this Monograph

Dr N. van Lookeren Campagne
Van Bergenlaan 6
2242 PV Wassenaar
The Netherlands
(Former Manager, Environmental Affairs, Shell Nederland B. V.)

Prof. J. C. I. Dooge
Centre for Water Resources Research
Civil Engineering Dpt.
University College Dublin
Earlsfort Terrace
Dublin 2, Ireland
(President of ICSU)

Ir. Joost van Kasteran
Pica
Laan van Oostenburg 45
2271 AN Voorburg
The Netherlands

Prof. P. S. Liss
IGBP Chairman's Office
School of Environmental Sciences
University of East Anglia
Norwich NR4 7TJ, U. K.
(Chair, IGBP)

Dr G. A. McBean
Assistant Deputy Minister
Atmospheric Environment Service
4905 Dufferin St.
Downsview, Ont. M3H 5T4, Canada
(Past Chairman, WCRP Joint Scientific Committee)

*R. E. Munn, J. W. M. la Rivière and N. van Lookeren Campagne (eds), Policy Making in an Era of Global Environmental Change, 169–178.* 

Dr R. E. Munn
Institute for Environmental Studies
University of Toronto
Toronto, Ont., Canada M5S 3E8

Prof. J. W. M. la Rivière
International Institute for Infrastructural, Hydraulic and Environmental Engineering
Westvest 7
PO Box 3015, 2601 DA Delft
The Netherlands

Mr P. Timmerman
Institute for Environmental Studies
University of Toronto
Toronto, Ont., Canada M5S 3E8

Dr P. Williamson
IGBP Chairman's Office
School of Environmental Sciences
University of East Anglia
Norwich NR4 7TJ, U. K.
(Chair, IGBP)

## Addresses of the Main Bodies Active in the Field of Global Environmental Change

Commission on Sustainable Development
c/o Undersecretary General N. Desai
Department of Policy Coordination and Sustainable Development
United Nations, NY 10017, USA

Earth Council
PO Box 2323-1002
San José, Costa Rica

GCOS
c/o WMO
Case postale 2300
1211 Geneva 2, Switzerland

GOOS
c/o IOC, UNESCO
7 Place de Fontenoy
75352 Paris, France

GTOS
c/o FAO
Viale delle Terme di Caracalla
00100 Rome, Italy

ICC
38 Cour Albert I
75008 Paris, France

ICSU
51 Blvd. de Montmorency
75016 Paris, France

IGBP Secretariat
Royal Swedish Academy of Sciences
Box 50005, Lilla Frescativägen 4
10405 Stockholm, Sweden

IHDP
11A Avenue de la Paix
1202 Geneva, Switzerland

IIASA
A-2361 Laxenburg
Austria

IOC
c/o UNESCO
7 Place de Fontenoy
75352 Paris 7, France

ISSC
1 Rue Miollis
75732 Paris 15, France

IUBS
51 Blvd. de Montmorency
75016 Paris, France

IUCN
28 Rue de Mauverney
1196 Gland, Switzerland

SCOPE
51 Blvd. de Montmorency
75016 Paris, France

START
Suite 200
2000 Florida Ave. N. W.
Washington, D. C. 20009, USA

UNEP
Box 30552, Nairobi, Kenya

UNEP Office for Industry and Environment
39–43 Quai André Citroën
75739 Paris Cedex 15, France

UNESCO
7 Place de Fontenoy
75352 Paris 7, France

WCRP
c/o WMO
Case postale 2300
1211 Geneva 2, Switzerland

WMO
Case postale 2300
1211 Geneva 2, Switzerland

World Business Council
160 Route de Florissant
1231 Conges
Geneva, Switzerland

## Address of the International Fund for Global Change Research

Foundation International Fund for Global Change Research (est. 1994), c/o ICSU, 51 Blvd. de Montmorency, 75016 Paris, France.

*Purposes*:
(1) Financial support for scientific programmes and projects studying Global Change in which ICSU participates.
(2) Fund-raising in all parts of the world and the investment of funds to support the aforementioned programs and projects.

*Bank account*: 48.57.67.67.317 with ABN/AMRO Bank, Oegstgeest, The Netherlands

## Addresses and Information on 14 Dutch Businesses* that Assisted Financially in the 1992 Symposium (see Preface by the Sponsors), and the Preparation of this Book (Alphabetically)

1. Akzo Nobel NV
2. AVEBE BA
3. BSO/Origin BV
4. CIVI Foundation
5. DSM NV
6. ECN Foundation
7. GE Plastics Europe
8. Koninklijke Hoogovens NV
9. Koninklijke KNP BT
10. Koninklijke Nedlloyd NV
11. NV SEP, electricity board
12. Shell Nederland BV
13. TNO, research organization
14. Vredestein NV

1. *Akzo Nobel NV* is an international company. Its core business is chemicals, coatings, fibres and health care. Its turnover in 1994 of US$ 13 000 million (DFL 22 200 million) 93% of which was realized outside the Netherlands. Akzo Nobel employs 70 400 people in a great many countries, 73% of them working outside the Netherlands.

   The concern for health, safety and environmental issues forms an integral part of Akzo Nobel's business policy. Akzo Nobel actively supports the guiding principles of the Business Charter for Sustainable Development of the ICC and the Responsible Care programme of the chemical industry.

   Akzo Nobel protects the environment by preventing or reducing the environmental impact of its activities and its products through appropriate design, manufacturing distribution use and disposal practices.
   The address of Akzo Nobel NV is Velperweg 76, 6800 Arnhem, the Netherlands.
   Tel. ++31.26.3664433; fax ++31.26.3663250.

2. *AVEBE BA* is a company whose core business is the production and sale of potato-, tapioca-, corn- and wheat-starches and their derivatives. Its turnover is DFL 1400 million and it exports 88% of its products. AVEBE employs 2600 people of whom 1000 work outside the Netherlands.

   Within the organization there is a 'Common Service and Total Quality Management Group'. It takes care of all aspects of quality control, environment and working conditions, in order to comply with the high standards of the company's 'grand strategy'. These standards are derived from the regulations of the countries in which the company operates.
   AVEBE's address is: PO Box 15, 9640 AA Veendam, the Netherlands.
   Tel. ++31.598.664286; fax ++31.598.664230.

* No information received from two other sponsors: the chemical company Hoechst Holland BV and the agricultural bank RABO BANK NED. BV.

3. BSO/Origin BV is an international full-service supplier in the area of information and communication technology. The turnover of the company in 1994 was DFL 816 million; 50.1% was realized outside the Netherlands. Employees total 5610; 3200 (57%) work outside the Netherlands.

   'BSO/Origin's mission regarding environment is to execute our business while making a positive contribution to the social and economic environment and minimizing negative impacts on the planet's already scarce resources. In the last few years, BSO/Origin has been maintaining an environmental account to measure the extracted value, the theoretical costs of repairing the cleaning of the damage caused by the company.'

   Address of the main office: BSO/Origin BV, PO Box 8549, 3503 RM Utrecht, the Netherlands.
   Tel. ++31.30.2586800; fax ++31.30.2586710.

4. *The Central Institute for Industry, CIVI* is a Dutch foundation. One of its present projects is to gauge the market for tailor-made answers to Earth System Research-related questions coming from business, governments and research establishments.

   CIVI has started these activities in the Netherlands, but looks for an international network of similar science-based organizations, which also want to provide these services. Good contact with a vast network of the world's relevant scientists is essential. See also in Chapter 8, 'Science-based consultancies'.

   CIVI's address is: Vlietweg 17, 2260 AD Leidschendam, the Netherlands.
   Tel. ++31.70.3201173; fax ++31.70.3177325.

5. *DSM NV* is an international company with its core business in base chemicals, base materials, fine chemicals, performance materials and materials processing. Its turnover in 1994 was DFL 8977 million, of which 41% was realized outside the Netherlands. DSM employs 19 113 people, 42% of whom work abroad.

   'In the area of safety, health and the environment, DSM aims at the very best. Our corporate objectives are: no accidents, no health problems attributable to the activities of DSM and no undue burdening of the environment. We are well aware that these objectives can be achieved only through a considerable and sustainable effort on our part.

   We are willing to make this effort, taking into account technological, organizational and financial constraints. In doing so we shall apply uniform standards throughout the world and shall at all times comply with local regulations.'

   The address of the main office is: DSM NV, het Overloon 1, 6411 TE Heerlen, the Netherlands (PO Box 6500, 6401 JH Heerlen).
   Tel. ++31.45.5778811; fax ++31.45.5719753.

6. *ECN*, the Netherlands Energy Research Foundation, is described by its President Prof. Ir. H. H. van den Kroonenberg in a section of Chapter 8, 'Designing energy systems for a sustainable future'.
   Its address is: ECN, Westerduinweg 3, 1755 LE Petten, the Netherlands (PO Box 1, 1755 ZG Petten).
   Tel. ++31.22456.4949; fax ++31.22456.4480.

7. *GE Plastics Europe* is a major chemical company in the Netherlands. The core business is the development, marketing and sales of engineering plastics; application areas: automotive, electrical/electronics, and building and construction. Employees total 3000; 1000 working outside the Netherlands.
   'European health, safety and environmental policy: it is the objective of GE Plastics Europe to manufacture and market engineering plastics that meet the highest quality standards, in facilities that meet similar environmental and safety standards. We will operate our plants and conduct our other activities in such a way that the safety and health of our employees, contractors, visitors and the surrounding community will not be exposed to unacceptable (as defined by our permits and state-of-the-art industry practices) risks. Safety and environmental protection is primarily a management responsibility, but it is also the responsibility of each individual GEP employee. Creating and maintaining safe and environmentally sound working conditions is an integral part of our business. This not only helps ensure the well-being of our personnel and the communities in which we operate but also contributes to total business success. In doing our jobs nothing is more important than meeting or exceeding our safety and environmental objectives.'
   Address of the main office: GE Plastics Europe, PO Box 117, 4600 AC Bergen op Zoom, the Netherlands.
   Tel. ++31.164.292225; fax ++31.164.291865.

8. *Koninklijke Hoogovens NV* are the main steelmills in the Netherlands and a large operator in alumina. Core business: the production of high-quality steel and aluminium, and in particular the preparation of finished products, rolled and extruded products and semi-finished products from these basic materials. In addition, the group is active in fields which have a clear connection with the core activities, such as steel processing and trading and the provision of technical services. The group's main premises are in the Netherlands, Germany and Belgium. The turnover of the company in 1994 was DFL 7934 million; 80% was realized outside the Netherlands. Hoogovens employs in total 19 000 people; 25% are working outside the Netherlands.
   As Hoogovens feels highly responsible for the environment, the company is striving continuously to reduce pollution per individual product by energy saving, reducing harmful pollutants, limiting the use of raw materials as far as possible and closed-loop recycling of both steel and aluminium.

Hoogovens' former president, Ir. OHA. van Royen, gives his views on 'Global research contributions to corporate strategy', in a section of Chapter 8 of this monograph.
Address of the main office: Koninklijke Hoogovens NV, P.O. Box 10000, 1970 CA IJmuiden, the Netherlands.
Tel. ++31.251.499111; fax ++31.251.470000.

9. *Koninklijke KNP BT* is a distribution, packaging and paper group. It produces paper, board and packaging materials. It is active in the distribution of paper, graphic and information systems. Sales in 1994 were approx. DFL 13 000 million; it employs more than 28 000 people. KNP BT's environmental policy is an essential part of management's responsibility. Starting from the existing and expected legal obligations, the company's policy is directed towards permanent improvement and development of processes and products. The general goal is to minimize the consumption of energy and raw materials in the operating companies.
The address of the main seat is: Koninklijke KNP BT., Paalbergweg 2, PO Box 23456, 1100 WZ Amsterdam, the Netherlands.
Tel. ++31.20.5672716; fax ++31.20.5672576

10. *Koninklijke Nedlloyd NV* is a major transportation group in the Netherlands. The core business is container logistics through a network of global shipping links, transport, forwarding, stock management and distribution, primarily in Europe. The turnover of the company in 1994 was DFL 6.6 billion; 77% was realized outside the Netherlands.
Employees total ± 19 000; 50% work outside the Netherlands.
Nedlloyd believes that efficient transport serves the interest of both the customer and the environment. Logistic know-how, hi-tech planning systems and optimum use of transport networks contribute to high-quality services at a competitive price *and* are the best guarantee for environmental protection. An environmental policy has been developed for European transport and distribution; sea transport; offshore drilling and production; transport equipment; and research and development. Nedlloyd does not only take into account international legislation, but also local law. This legislation has an impact on, for instance, the means of transport and the cargo. As far as national and international environmental plans are concerned, Nedlloyd not only closely follows the debate, but also tries to provide valuable contributions.
Address of the main seat: Koninklijke Nedlloyd NV, PO Box 487, 3000 AL Rotterdam, the Netherlands.
Tel. ++31.10.4007111; fax ++31.10.4006460.

11. *NV SEP*, the Dutch electricity generating board, is a collaborative alliance of the four regional power companies generating electricity and producing heat on a large

scale for public supply. The general philosophy of this board is explained by its President Ir. N G. Ketting in a section of Chapter 8, 'Agreement on long term goals necessary for sustainability'. SEP has been a sponsor for years of the ICSU International Geosphere Biosphere Program of Chapter 3 of this monograph.
Address of main office: NV SEP, Utrechtseweg 310, 6812 AR Arnhem, PO Box 575, NL-6800 AN Arnhem, the Netherlands.
Tel. ++31.26.3721111; fax ++31.26.4430858.

12. *Shell Nederland BV* is a part of the Royal Dutch/Shell Group of companies, one of the largest businesses in the world. Its core business is oil, gas and chemicals. Shell Nederland environmental policy guidelines are: progressive reduction of emissions, effluents and the discharges of waste materials that are known to have a negative impact on the environment, with the ultimate aim of eliminating them.
Shell Nederland sponsored the International Geosphere Biosphere Program during its first three years of development. In a section of Chapter 8, 'Research can help to avoid conflicts of interest', Ir. J M H. van Engelshoven, a former board member of the Royal Dutch/Shell Group of Companies, gives some of his views on global change.
Address of the main seat is: Shell Nederland BV, Shell-gebouw, Hofplein 20, 3032 AC Rotterdam, the Netherlands.
Tel. ++31.10.4696911; fax ++31.10.4116828.

13. *TNO*, the Netherlands organization for applied scientific research, is committed to the Earth's sustainable development. It has an institute dedicated to environmental matters – TNO Institute of Environmental Sciences, Energy Research and Process Innovation. TNO has also played an important role in the establishment of the Environmental Technology Valley. This initiative by Dutch universities, businesses, intermediary organizations and TNO caters to fundamental know-how development, applied research, implementation of environmentally related products and processes. Its core business is the application of scientific research and technological know-how in the fields of the environment, industrial technology, agriculture/food, defence, health care and technology policies. TNO's turnover is US$ 406 million (DFL 650 million) of which 14.8% is realized abroad. It has 4100 employees, mainly working inside the Netherlands.
Address: c/o TNO Corporate Communications Dept., P.O. Box 6050, 2600 JA Delft.
Tel. ++31.15.2694990; fax ++31.15.2627335; E-mail infodesk@nl.tno.

14. *Vredestein NV* is a holding company in a decentralized organization, in which group companies are responsible for their own operations and results. Its core business is: production, sale and distribution of rubber and synthetic products, such as tyres, tubes, boots, recycled rubber, etc. Turnover in 1994 was DFL 453 million,

with 80% realized outside the Netherlands. It employs about 2000 people of which 10% work outside the Netherlands.

Further substantial strategic investments are to be made, particularly at Vredestein Banden. This investment programme includes plans to raise quality, flexibility and productivity to an even higher level. The investment expenditures will be roughly equal to the cash flow. Measures to achieve further improvements in productivity and cost controls will continue with undiminished vigour.

Address of the main office: Vredestein NV, P. O. Box 200, 6880 AE Velp, the Netherlands.
Tel. ++31.26.3657911; fax ++31.26.3648639.

## Closing Remarks

These 16 Dutch companies (including Rabo bank and Hoechst Holland) have in common that a predominant part of their turnover is realized outside the Netherlands, and exports play a large role. It shows the very open economy of the Netherlands and the international outlook. Also about half of the employees of these companies work and live abroad. This explains the interest in international developments, including research programmes that increase our understanding of the Earth System, with its many phenomena which have a direct long-term impact on business.

# Appendix 2. List of Acronyms

| | |
|---|---|
| ACSyS: | Arctic Climate System Study |
| APN: | Asian Pacific Network for Global Change Research |
| ASCEND 21: | An Agenda of Science for Environment and Development into the 21st Century |
| BAHC: | Biospheric Aspects of the Hydrological Cycle |
| CEA: | Cumulative environmental assessment |
| CFCs: | Chlorofluorocarbons |
| CLIVAR: | Climate Variability and Predictability |
| DIS: | Data and Information System |
| DMS: | Dimethylsulphide |
| EC: | European Commission |
| EIA: | Environmental impact assessment |
| ENGO: | Environmental Non-Governmental Organization |
| ENSO: | El Niño-Southern Oscillation |
| EU: | European Union |
| GAIM: | Global Analysis, Interpretation and Modelling |
| GAW: | Global Atmospheric Watch |
| GCM: | General circulation model |
| GCOS: | Global Climate Observing System |
| GCTE: | Global Change and Terrestrial Ecosystems |
| GEWEX: | Global Energy and Water Cycle Experiment |
| GRIP: | Greenland Ice-core Project |
| Gt: | Gigatonne |
| GTOS: | Global Terrestrial Observing System |
| GEF: | Global Environmental Facility |
| GEMS: | Global Environmental Monitoring System |
| GLOBEC: | Global Ocean Ecosystem Dynamics |
| GOOS: | Global Ocean Observing System |
| IAI: | InterAmerican Institute for Global Change Research |
| ICC | International Chamber of Commerce |
| ICSU: | International Council of Scientific Unions |
| IDNDR: | International Decade on Natural Disaster Reduction |
| IGAC: | International Global Atmospheric Chemistry Project |

*R. E. Munn, J. W. M. la Rivière and N. van Lookeren Campagne (eds), Policy Making in an Era of Global Environmental Change, 179–180.* 

IGBP: International Geosphere–Biosphere Programme
IGFA: International Group of Funding Agencies for Global Change Research
IHDP: International Human Dimensions of Global Environmental Change Programme
IHE: International Institute for Infrastructural, Hydraulic and Environmental Engineering
IIASA: International Institute for Applied Systems Analysis
IOC: Intergovernmental Oceanographic Commission
IPCC: Intergovernmental Panel on Climate Change
ISSC: International Social Science Council
IUBS: International Union of Biological Sciences
IUCN: World Conservation Union
JGOFS: Joint Global Ocean Flux Study
LOICZ: Land–Ocean Interactions in the Coastal Zone
LUCC: Land Use and Land Cover
MAB: Man and the Biosphere Programme
MINK: Missouri, Iowa, Nebraska, Kansas
NAPAP: U.S. National Air Pollution Assessment Program
NGO: Non-Governmental Organization
OECD: Organization for Economic Cooperation and Development
PAGES: Past Global Changes
PPP: The Polluter Pays Principle
SCOPE: Scientific Committee on Problems of the Environment
SPARC: Stratospheric Processes and their Role in Climate
START: Global Change System for Analysis, Research and Training
TOGA: Tropical Oceans and the Global Atmosphere
UNCED: United Nations Conference on Environment and Development
UNEP: United Nations Environment Programme
UNESCO: United Nations Educational, Scientific and Cultural Organization
UV-B: Ultraviolet-B radiation
WCRP: World Climate Research Programme
WICEM: World Industry Conference on Environmental Management
WMO: World Meteorological Organization
WOCE: World Ocean Circulation Experiment

# Appendix 3. Methods for Assessing Effects of Global Change on the Biosphere and Society

R. E. MUNN
*Institute for Environmental Studies, University of Toronto, Canada*

## Methods Available

Many studies suggest that future changes in the state of the global environment will be large, and will occur in the 21st century. If this projection is realized, the biosphere will be greatly stressed, to the point that many ecosystems may have to adapt profoundly to new sets of environmental conditions, or collapse. This Appendix deals with the methods that are used to estimate the effects of global change on the biosphere and society. (No attempt is made to describe the effects themselves; the literature on that subject is voluminous, and in most cases is given as likely answers to sets of 'what-if' questions. The nature, timing and severity of effects depend on assumptions about the strengths of the socioeconomic driving forces.)

Over the last two decades, scientists have been attempting to identify and quantify the impacts of global change, particularly climate change, on the biosphere and society. Three general approaches are taken:

(1) *Sectoral studies*, especially with respect to managed systems (e.g., agriculture, forestry) and with respect to a single type of global change (e.g., climate change or demographic change); see p. 182.
(2) *Integrated studies* across a range of sectors and types of global change, in which interactions are included; see p. 184.
(3) *Whole ecosystem studies*, applying some relatively recent ideas variously referred to as 'ecosystem integrity', 'resilience', 'robustness' and 'the ecosystem approach'. See p. 186. The idea of an ecosystem as a self-organizing system that survives by continual evolution brings a new perspective to the question of managing global change.

No matter which approach is taken, there is real concern amongst specialists that severe impacts could occur.

*R. E. Munn, J. W. M. la Rivière and N. van Lookeren Campagne (eds), Policy Making in an Era of Global Environmental Change, 181–188.* 

A climate change sufficient to have noticeable impacts on the economy of a region will take time, probably several decades at least. In the modern world, population growth and migration, *per capita* income growth, technological change in production and consumption, shifting consumer preferences across the vast spectrum of traditional and newly available goods and services, and changes in relative power and influence among regions and nations mean that the social, cultural, and economic structure of any region almost certainly will change significantly over a period of several decades

Pierre R. Crosson and Norman J. Rosenberg (1993).

## Sectoral Impacts of Global Change

Most global-change impact studies are sectoral in nature, and are based on a combination of:

(1) Socioeconomic scenarios, usually national or regional in scale (except in the case of greenhouse gas projections, which are generally global in scale).
(2) Global climate change scenarios obtained from historical analogues or from general circulation models, together with assumptions about rates of increase of the greenhouse gases;
(3) Global biosphere models, in which biosphere responses are coupled to global climate change scenarios. (The biosphere includes people, so these models include, for example, human health components.)
(4) Models of second-order effects, e.g., assessments of reductions in food production on world trade patterns.

One of the problems with such studies is that of 'downscaling' from global to regional simulations (Frey-Buness *et al.*, 1995).

Here we present only a few examples of these studies. For general reviews, see Gates (1993), Kareiva *et al.* (1993), Frederick and Rosenberg (1994) and IPCC (1996). This is an area of very considerable effort at the present time.

### *Impacts of Climate Change on World Agriculture*

Martin Parry (1990) and associates have examined the effects of climate change on world agriculture. They have used a conceptual framework consisting of:

(1) A range of models relating to: (a) climate change; (b) direct relationships between climatic variables and plant growth/crop yield/rangeland carrying capacity; (c) second-order relationships between yields and production/employment/profitability at the farm level; higher-order relationships between regional/national agricultural yields and agricultural employment/activity rates in non-agricultural sectors/etc.

(2) Identification of changes caused by ecologically-related changes, e.g., changes in rates of soil erosion, pest and disease outbreaks, groundwater depletion and acid deposition patterns.
(3) Identification of technical adjustments by the farmer and regional/national policy responses.

The weakest link in this assessment chain is the very first one. Climate change models are inadequate at the regional level. Use is therefore made of critical *thresholds*, where ecological and socioeconomic responses to global change would become measurable and/or significant. This approach is supplemented by a sensitivity study designed to identify the climate elements to which crops are particularly sensitive, and to quantify threshold changes in the climatic elements that would seriously impact crop yields (or on the other hand, improve productivity, like increasing the frost-free season).

For an overview of other studies on the impacts of climate change on agriculture and forestry, see WMO (1994).

*Geographic Changes in Global Vegetation Patterns Due to Climate Warming*

Monserud *et al.* (1993) have presented the results of some simulations of the effects of climate warming on the geographic distribution of world vegetation belts. Using as a scenario the case of doubled greenhouse gas concentrations from their preindustrial levels, climate models predict that warming will be greatest in polar latitudes, and least near the equator. As expected, climate–vegetation models therefore predict that the greatest changes in the vegetation belts will take place in the boreal and temperate zones. In particular, all boreal class types of vegetation are expected to shrink. Most vegetation classes

**Box 1**

The main sectors usually considered in studies of biosphere and socioeconomic responses to global change:

- Agriculture
- Forestry
- Unmanaged terrestrial ecosystems
- Permafrost
- Energy
- Water resources
- Fisheries
- Coastal zones affected by sea level rise
- Industry and transportation
- Human settlements
- Human health
- Financial (insurance, costs of environmental refugees, etc.)

in the subtropics and tropics are predicted to expand, but this result is uncertain because of the difficulty in modelling changes in rainfall. The most stable areas are desert and ice/polar desert.

Here it should be mentioned that adjustments to the world's vegetation belts may lag behind climate changes by up to 50–100 years, and in some cases the new patterns may be rather distorted because of unsuitable soils for colonization, into the tundra, for example, of a particular vegetation type.

### *Impact of Climate Change on Malaria Vectors*

Climate is of great significance for both endemic and epidemic malaria, and for the survival of mosquito vectors. Theo Jetten and Willem Takken (1994) have reported on a study of the effects of climate change on malaria vectors in Europe, using simulation models. Preliminary results suggest that a temperature increase of 2 deg C would indeed increase the numbers of infectious mosquitoes, particularly in the Mediterranean region, where there could be a 100-fold increase in infectious mosquito populations.

## Integrated Assessments

### *The Use of Historical Analogues*

Sectoral assessments generally overlook interactions amongst sectors. A much richer range of scenarios is provided by integrated assessments, in which attempts are made to capture these interactions. Of course, the deceptively easiest way to undertake an integrated assessment is to examine an historical analogue – an extremely hot summer, an extremely cold winter, or a prolonged drought, for example. Such assessments provide clues but cannot reproduce exact analogues. In particular, the biosphere, or society, may be able to withstand an environmental stress lasting a season or even a decade. For sustained change, system resilience might be severely overtaxed.

A good example of the analogue approach is the study by Salafsky (1994) of the impacts of a drought on the rural economy in West Kalimantan in Indonesia. In this rainforest environment, a drought in 1991 caused severe economic hardship to the residents of several small villages studied. This arose because of reduced fruit harvests, losses of coffee plantations, delays in rice crops, increased efforts expended in obtaining drinking water, lost wages in the forest product sector, and increased health problems. The *per-capita* loss in income was estimated to be in the range of 23–47%.

This case study provides some indication of what might be the economic consequences if drought frequencies increased in this region. It also helps in the development of policies to protect against such eventualities, e.g., through construction of improved irrigation and household water distribution systems.

Another example of the climate analogue approach is the widely referenced 'MINK' study, which is an integrated climate change impact assessment of four states in the American mid-west: *M*issouri, *I*owa, *N*ebraska and *K*ansas (Rosenberg, 1993). Because of uncertainty in climate change simulations on a regional scale, the investigators chose to use climate analogues based on historical meteorological and stream-flow data from the drought years of the 1930s. Recognizing, however, that future socioeconomic conditions may be quite different from present ones, the climate analogues were applied to socioeconomic scenarios for the year 2030 rather than present-day ones. Another innovative feature of the study was recognition that people adversely affected by a drought or flood would not suffer their fates passively; instead, the investigators considered how economic and technological change by the year 2030 would affect sensitivity to climate, and how adaptation would alter that sensitivity. Finally, the study examined the effect of increased carbon dioxide concentrations on increased photosynthesis and water use efficiency by vegetation.

*The Mackenzie River Basin Study: an integrated assessment based on scenarios*

The Mackenzie river basin is a large watershed in northwestern Canada just east of the Rocky Mountains. This is the subject of a current study whose main objective is to assess the potential impacts of global warming and socioeconomic development on the region (MBIS, 1994; Cohen, 1995). Although only a description of 'work in progress' can be given (no detailed results are expected before 1998), the study is included here because the methodologies used represent a major step forward.

An integrated assessment approach is being employed, designed to answer the following questions:

(1) What are the implications of climatic change for achieving regional resource development objectives? Should governments within the Mackenzie basin alter their current resource-use policies or plans regarding water resources, resource extraction, forests or fish and other wildlife in anticipation of global warming?

(2) Does climatic change increase land-use conflicts among different economic and social sectors? If potential conflicts are identified, how serious might they be and how could compromises be reached?

(3) What are the possible tradeoffs for alternative public responses to climatic change? Should parks and forests be managed to anticipate change, or to preserve existing conditions? What are the implications for fire control, recreation, tourism and wildlife management?

(4) What are the implications of global warming for community management of resources under land claims agreements?

Three scenarios of warmer climates for the year 2050 have been obtained from general circulation models, and a fourth one has been derived from historical climate analogues.

A population growth model and an input–output economic model have been used to obtain four socioeconomic development scenarios. Work is now in progress to obtain the necessary databanks to test model performances and to permit synthesis of the various sectoral results.

An important element in study design has been the involvement of community groups, including native people, to ensure that the ecosystem components selected for study are relevant not only to scientists but also to the population of the region. In this connection, Project leaders are searching for ways to integrate the results obtained from 28 freshwater and terrestrial studies with aboriginal knowledge.

### *Socioeconomic Impact of Sea Level Rise in the Netherlands*

Den Elzen and Rotmans (1992) have developed four sets of socioeconomic scenarios for the Netherlands based on differing economic growth, energy use, international environmental actions, etc. These have been applied to a climate-change model called IMAGE (Alcamo, 1994), which was run over the period 1900–2100 to provide estimates of sea level rise. The socioeconomic costs of these rises were then estimated:

(1) By the year 2100, the dykes will have to be raised by about 2 m, which would cost about 20 billion guilders.
(2) In addition to raising the dykes, it would be prudent to intensify studies of ways to stop coastal dune erosion. Implementation of such measures would cost an additional 5–6 billion guilders over the next century.
(3) In addition, salt water seepage into IJssel Lake would have to be prevented, at a further cost of 10 billion guilders by the year 2100.

These economic costs of sea level rise in the Netherlands should be viewed in a risk assessment framework. That is, given a sea level rise of, say, 2 m over the next hundred years, the cost of protecting the coastline from major socioeconomic impacts would be about 35 billion guilders. This value then should be compared with current estimates of the Dutch share of the cost of reducing greenhouse gas emissions worldwide.

## Applying Some Modern Ideas on Ecosystem Integrity to Global Change Assessments

Ecosystems are not in a steady state, as assumed in many classical models, e.g., that of climax vegetation. Instead, ecosystems are constantly evolving, 'hunting' for, but never achieving, equilibrium. This is because externalities are changing too – on time scales ranging from a few days to millennia.

Environmental management policies are often based on the principle that an ecosystem should be preserved in its present form. This eventually turns the system into a relict of the past, and it loses its ability to cope with change. Ecosystems that are most likely to adapt successfully to external changes have a self-organizing ability, which is referred to as *ecosystem integrity* (Edwards and Regier, 1990; Kay, 1991). The mathematics required to investigate this 'new' ecology is complex, involving non-linear differential equations and the theory of chaos. But it is likely that in coming decades, some of the ideas involved will influence management options with respect to the effects of global change on ecosystems and society. As an example, the approach to preserving ecosystem integrity is based on the need to conserve ecosystem processes/functions rather than individual species, particularly in a changing world.

## Concluding Remarks

Most studies on the effects of global change on the biosphere and society are undertaken in individual research institutes. Many of the results are reported at national and international meetings and ultimately appear in peer-reviewed journals. The role of intergovernmental and non-governmental bodies has largely been to promote international syntheses of current knowledge and to identify knowledge gaps. The following are examples:

(1) The WMO recently published a Technical Note (No. 196): *Climate Variability, Agriculture and Forestry* (1994), which contains substantial sections on impacts of climate change.
(2) SCOPE (Scientific Committee on Problems of the Environment) has published several synthesis volumes relating to the impacts of global change on the biosphere, and has three more in preparation.
(3) The IPCC (Intergovernmental Panel on Climate Change) has recently published some technical guidelines for assessing climate change impacts and adaptations (IPCC, 1994);
(4) The IPCC (1996) has published a monograph on the effects of climate change on the biosphere.

These syntheses are very useful, not only in influencing research directions, but also in contributing to policy analyses.

## Selected References

Alcamo, J. (ed.) (1994) *IMAGE 2.0: Integrated Modelling of Global Climate Change*. Kluwer, Dordrecht, The Netherlands.
Cohen, S. (1995) Integrated assessment of long-term climatic change. In: *Looking Ahead: The Inclusion of Long-Term Global Futures in Cumulative Environmental Assessments* (R. E. Munn, ed.), Environmental Monograph No. 11, Institute for Environmental Studies, University of Toronto, pp. 115–128.

Crosson, P. R. and Rosenberg, N. J. (1993) An overview of the MINK study. *Climatic Change*, **24**, 159–173.

Den Elzen, M. G. J. and Rotmans, J. (1992) The socio-economic impact of sea-level rise on the Netherlands: a study of possible scenarios. *Climatic Change*, **20**, 169–195.

Edwards, C. J. and Regier, H. A. (1990) *An Ecosystem Approach to the Integrity of the Great Lakes in Turbulent Times*. Great Lakes Fisheries Commission, Special Publication 90–4, 1452 Green Road, Ann Arbor, MI; 299 pp.

Frederick, K. and Rosenberg, N. J. (eds) (1994) Assessing the impacts of climate change on natural resource systems. *Climatic Change*, **28**, 1–219 (also published as a book by Kluwer).

Frey-Buness, F., Heimann, D. and Sausen, R. (1995) A statistical–dynamical downscaling procedure for global climate simulations. *Theor. Appl. Climatol.*, **50**, 117–131.

Gates, D. M. (1993) *Climate Change and its Biological Consequences*. Sinauer Associates, Sunderland, MA, 280 pp.

IPCC (1994) *IPCC Technical Guidelines for Assessing Climate Change Impacts and Adaptations* (T. R. Carter, M. L. Parry, H. Harasaw and S. Nishioka, eds). IPCC/WMO/UNEP, CGER-1015-'94, Geneva, 59 pp.

IPCC (1996) Report of Working Group II to the IPCC Second Assessment Report, WMO/UNEP Geneva (in press).

Jetten, T. H. and Takken, W. (1994) Impact of climate change on malaria vectors. *Climatic Change*, **18**, 10–12.

Kareiva, P. M., Kingsolver, J. G. and Huey, R. B. (eds) (1993) *Biotic Interactions and Global Change*. Sinauer Associates, Sunderland, MA, 559 pp.

Kay, J. J. (1991) The concept of ecological integrity, alternative theories of ecology, and implications for decision-support indicators. In *Economic, Ecological and Decision Theories*. Canadian Environmental Advisory Council, Environment Canada, Ottawa, pp. 23–49.

MBIS (1994) The Mackenzie Basin Impact Study: Interim Report No. 2, Environment Canada, Room 210, Twin Atria no. 2, 4999–98 Ave, Edmonton T6B 2X3, 485 pp. (see also Newsletters nos 1–4).

Monserud, R. A., Tchebakova N. M. and Leemans, R. (1993) Global vegetation change predicted by the modified Budyko model. *Climatic Change*, **25**, 59–83.

Parry, M. (1990) *Climate Change and World Agriculture*. Earthscan, London.

Rosenberg, N. J. (ed.) (1993) Towards an integrated impact assessment of climate change: the MINK study. *Climatic Change*, **24**, 1–173.

Salafsky, N. (1994) Drought in the rain forest: effects of the 1991 El Nino-Southern Oscillation event on a rural economy in West Kalimantan, Indonesia. *Climatic Change*, **27**, 373–396.

WMO (1994) Climate Variability, Agriculture and Forestry. Technical Note No. 196, WMO.

# Appendix 4. Instruments for Facilitating the Application of Global Change Research Results to Policy-Making

R. E. MUNN
*Institute for Environmental Studies, University of Toronto, Canada*

## Consensus-Building

The scientific results of global change studies must be presented to stake-holders in a policy-relevant fashion if they are to respond in meaningful ways. The principal stake-holders are:

- Governmental and intergovernmental bodies who have responsibility for environmental policy development.
- The public, whose views are reflected through their elected representatives, as well as through trade unions and environmental groups such as the Sierra Club and Greenpeace.
- Business and industry, whose views are frequently channelled through organizations such as Chambers of Commerce;
- The scientific community, whose views are reflected through Academies of Science, National Research Councils, single-discipline scientific societies, and national panels and task forces established specifically to seek a consensus on a particularly troubling issue.

Some of the tools used to achieve consensus on actions to be taken with respect to environmental issues are:

(1) *Informing the media and the public interest groups*
Some countries and some intergovernmental bodies (e.g., OECD, UNEP) publish State-of-the-Environment (SOE) reports, providing useful information on trends in the quality of the environment. These reports are sometimes too expensive for citizens to buy, but there are usually supplementary channels of communication through Fact Sheets on specific issues, and information booths at agricultural and other fairs. Science writers and TV documentaries also play an important role.

(2) *Consensus-building through scientific panels*
Particularly in the case of emerging new issues, governments often establish 'blue-ribbon' scientific Panels or Royal Commissions. The resulting reports

*R. E. Munn, J. W. M. la Rivière and N. van Lookeren Campagne (eds), Policy Making in an Era of Global Environmental Change, 189–196.* 

provide a basis for national debates. These Panels are composed of specialists and thus help to clarify the issues and to establish research agendas, although they are open to the criticism that their recommendations do not adequately reflect the views of citizen groups. This mechanism has been particularly effective in Britain and the United States.

(3) *Round Tables*
The *Round Table* is a Canadian experiment that was initiated following publication of the Brundtland Report on environment and development. The Canadian government established Round Tables, both nationally and provincially, with representatives (at a very high level) of government, industry, business, the trade unions, the 'green' organizations, and the scientific community. These Round Tables appear to be working reasonably well in most cases as sounding boards for the various stakeholders concerned.

(4) *Public hearings*
In some countries, public hearings are effective in at least determining the range of opinions prevailing across the citizenry with respect to particular environmental issues. The Panel or in some cases a single judge then tries to prepare compromise recommendations that are as fair as possible to all stake-holders. In the UK, for example, the Royal Commission is quite an effective mechanism for achieving consensus, given a range of diverging views.

(5) *Land-use planning guidelines*
Although not always practised, it is helpful for government bodies to promulgate draft policies for public discussion well in advance of the preparation of a specific land-use proposal. For example, the goals of a land-use policy might be: (a) to conserve soil fertility over the long term; (b) to reduce the loss of arable land; (c) to apportion available land more appropriately. Without guidelines and standards, a developer has no way of knowing whether the environmental effects of the development will be acceptable or unacceptable in the jurisdiction involved. Another reason for promulgating guidelines and standards is to control small-scale cumulative development. In some areas, many people can recall from their childhood, the large strips of countryside that have been lost to urbanization.

## Environmental Impact Assessment (EIA) and Cumulative Environmental Assessment (CEA)

The EIA is a widely used mechanism to gain public input into the development process. The idea originated in the United States in 1970, and quickly spread around the world. The EIA was applied originally to proposals to construct large power stations, flood control systems, smelters and so on. An extension to policies and programmes (e.g., long-term energy supply/demand proposals, international trade, forest management) has not yet been so widely practised.

An *environmental impact assessment* (EIA) is a process designed to: identify the impacts of a proposed development (power station, smelter, highway, etc.); to determine the magnitude and probability of occurrence of these impacts; to estimate their significance; to recommend mitigative measures (e.g., alternative sites or engineering designs; payments to displaced people); and to synthesize, interpret and communicate the results, generally in the form of an *environmental impact statement* (EIS). In most jurisdictions, the EIS is open for public comment, or is a background document at a public hearing (SCOPE 5, 1979).

An EIA/EIS may also be required for class actions (e.g., specific types of development in a region) and sometimes to support new policies (changes in pesticide spraying regulations, international trade agreements, etc.).

The EIA is often a useful tool but it sometimes leads to confrontation rather than consensus – for several reasons:

(1) Low-probability high-impact risks are difficult to discuss at a public forum. The depth of local feeling about a planned toxic waste facility, for example, cannot be overestimated – this is the familiar NIMBY (not-in-my-backyard) syndrome, sometimes called LULU (locally unacceptable land use).

(2) Intervenors at a public hearing concerning a specific development proposal may use the opportunity to debate broader issues, e.g., national energy policies rather than the environmental impacts of a proposed nuclear power station. On the other hand, public hearings that are couched in a legalistic framework may lead to frustration by citizens who have important points to make about these broader issues.

(3) Many EIAs (e.g., in the UK) deal with the possible negative impacts of a development but do not consider the socioeconomic benefits. Winners and losers are rarely identified, and benefits and disbenefits are rarely quantified.

(4) Because many developments are built to operate for 50 years or longer, the effects of global change ought to be included in an EIA.

Because of these problems the EIA process has not always been the unqualified success that its originators had expected. For example, a public hearing took place in Britain in the early 1990s to consider a proposal to build a motorway which would cross a medieval battlefield, Naseby. Writing in *The Times* (6 August 1993), Graham Searjeant noted that if a less adversarial approach to decision-making had been taken, no British planner would have ever suggested the proposed route for the motorway. He believes that public consultations should take place well in advance of the establishment of a Board to organize the preparation of an EIA.

Referring particularly to the cumulative effects of development, EIAs generally are prepared in isolation from what might happen in the surrounding region. Over the long term, however, three kinds of effects may occur:

(1) The development might cause slow environmental deterioration, e.g., acidification of a lake downwind of a smelter, or gradual on-site waste accumulation, which might take years before a problem developed. Yet intervenors at public hearings are often not permitted to discuss the possible long-term effects caused by the enterprise itself.
(2) The development might attract other significant enterprises to the region, causing additional impacts. For example, construction of a power station could make the surrounding area very desirable for locating a refinery or steel mill. Similarly the character of a region might change with the construction of roads, houses and shopping centres.
(3) Global changes might cause regional changes, with serious consequences for the development.

These effects are often likely to be important over the long term because many developments have life expectancies of 50–100 years, and in some cases make permanent commitments of natural resources, e.g., suburbia, reservoirs.

To overcome some of these difficulties, the idea of a 'cumulative environmental assessment' (CEA) has recently been promoted. In Canada, for example, the 1992 Canadian Environmental Assessment Act requires that cumulative effects of federally funded projects be examined.

CEA is a welcome addition to the EIA process but practitioners have generally been concerned only with cumulative effects of the first and second types described above. With respect to the third category, the inclusion of 'global change' in CEAs was the subject of a recent Workshop (Munn, 1994) which concluded that:

(1) CEAs must indeed take account of long-term global change.
(2) CEAs must therefore be carried out in the first instance at the regional or national level, and for policies, plans and programmes. CEAs for individual development proposals like power stations or highways could then be undertaken within this overall framework.
(3) Because the long-term is uncertain, the best approach is an adaptive one, in which the goal of a CEA is to explore the long-term consequences of alternative growth management strategies. In order for this to be useful as a policy tool, a supporting early-warning monitoring system must be put in place and continually updated.

## Establishing Long-term Environmental Priorities

With so many socioeconomic and environmental issues needing urgent attention, how are priorities to be established? How is society to decide, for example, on whether to spend money on greenhouse gas emission reductions, toxic waste disposal facilities, or development packages to economically poor countries? This is not an easy question to answer, but a few helpful suggestions can be mentioned.

### *Major Issues should be Considered in an Integrated Way*

Global issues are often treated one by one. However, many linkages exist, and in order to decide on priorities across a range of global issues, an integrated approach is necessary. For example, climate change, stratospheric ozone depletion, the usage of nitrogen fertilizers, deforestation and the management of rice paddies and wetlands are all interrelated through the global biogeochemical cycles. Although an integrated approach increases the complexity of the analysis, it may improve the policy relevance of the results (see Figure 1).

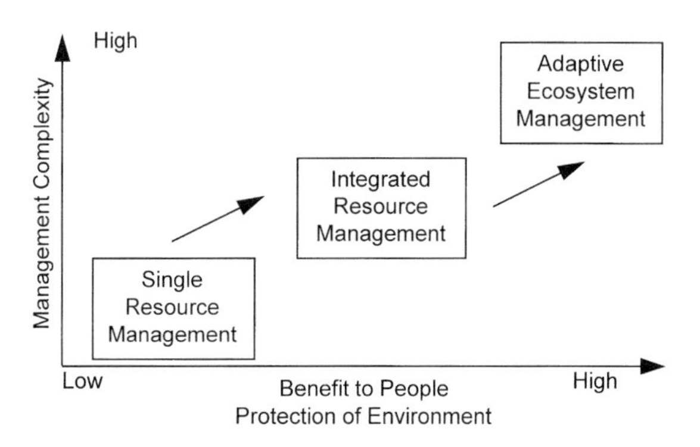


Fig. 1. Benefit of an integrated approach to environmental management (Ontario Forest Policy Panel, 1993)

A report of a recent workshop on multi-issue assessments has been published (Munn, 1995) and another organized by EC (1996) is in press.

### *The Need for Informal Consultations*

'Informal' consultation amongst stakeholders is a way of promoting international cooperation. Richard Benedick (1993) uses the term *policy dialogue*, which he recommends for several reasons:

(a) The informal process can establish personal rapport amongst various stakeholders and their technical advisors, freed from the constraints of having to represent particular constituencies.
(b) Issues can be broken down into manageable components, each of which may require further study to clarify various technical questions.

(c) Dialogue may plant the seed of an idea amongst participants. This idea can be explored informally.
(d) Networks can be developed amongst government officials, academics, and industrial and business leaders, who normally would not meet in such a round-table way.

Policy dialogues usually take place over 2–3-day periods in a 'retreat' setting.

### *Structured Forms of Policy Dialogues*

Highly structured forms of Benedick's policy dialogue have been developed, using various decision support systems such as gaming theory, backcasting and policy exercises. In many cases, the results of model simulations are displayed on video screens for various sets of input assumptions, permitting assessments of the consequences of several different policy options. Thus for example, comparisons can be made of the cost-effectiveness of various control strategies for reducing oxides of sulphur and nitrogen across Europe.

### *The IIASA* Policy Exercises*

In the late 1980s 'policy exercises' (for lack of a better phrase) were developed at IIASA. These used a very structured form of policy dialogues, which incorporated three innovations:

(1) In advance of a planned 2–3-day workshop, the 'home team' develops a range of possible 'futures', including some 'not-impossible' ones, without trying to estimate their probabilities of occurrence. These several scenarios are called 'future histories', and they must be internally consistent and believable. Because predicting the future becomes increasingly difficult as the time horizon lengthens to decades, the goal of scenario construction is an appropriate alternative. The scenarios are developed using simulation models enriched with a few surprises (e.g., a severe drought lasting several months; a technological break-through, making it possible to produce hardwood chemically; the discovery of an abundant new source for oil and gas).
(2) At the beginning of the workshop, participants are presented with a dilemma, involving uncertainties and socio-economic/environmental trade-offs (e.g., over-use of water from the Danube River basin with conflicts developing amongst the

* International Institute of Applied Systems Analysis, Laxenburg, Austria.

various stake-holders). The participants are provided with all relevant historical and current data, and are asked to design policies for the next 20–30 years.

(3) The home team then selects (randomly) one of the future histories, and workshop participants are invited to evaluate their policy recommendations in the light of these developments. The net result is usually a better appreciation of the need to pursue adaptive strategies, and to establish a set of leading socioeconomic and environmental early warning indicators.

In order to persuade senior-level people to participate in policy exercises they must be given real issues, with hard choices to be made involving winners and losers. In the IIASA European Futures study (Stigliani *et al.*, 1989), the policy exercises extended forward to the year 2030, when climate warming might become apparent. The dilemmas that were discussed included: water management, soil acidification, forest wood supply, marginalized land, sea level rise and pollution, chemical time bombs, non-point-source toxics, transport, urbanization and summer oxidant episodes. These were examined in the context of four socioeconomic pathways: present global trends continuing; higher growth rates around the world; environmentally friendly economic conditions around the world; and environmentally friendly economic conditions in Europe but not elsewhere. The results of this assessment are shown in Table 1. Summarizing briefly, environmental issues already of concern in the 1980s in Europe were soil acidification, transport and

Table 1. Ranking of dilemmas for the 1980s and according to development path for the year 2030. The numbers 1, 2, 3 represent degree of seriousness, from low to high

| Dilemma | 1980s | Pathway (1) Present trends continuing (Europe and elsewhere) | Pathway (2) High-growth economy/low environmental concern (Europe and elsewhere) | Pathway (3) Environmentally friendly economy (Europe and elsewhere) | Pathway (4) Environmentally friendly economy (Europe but not elsewhere) |
|---|---|---|---|---|---|
| Water management | 1 | 2 | 3 | 1 | 3 |
| Soil acidification | 2 | 3 | 3 | 1 | 1 |
| Forest wood supply | 1 | 2 | 3 | 1 | 3 |
| Marginalized land | 1 | 2 | 2 | 1 | 2 |
| Coastal issues | | | | | |
| Sea level | 1 | 2 | 3 | 1 | 3 |
| Pollution | 1 | 2 | 2 | 1 | 1 |
| Chemical time bombs | 1 | 2 | 3 | 1 | 2 |
| Non-point toxics | 1 | 2 | 3 | 1 | 1 |
| Transport growth | 2 | 3 | 3 | 2 | 2 |
| Urbanization | 1 | 1 | 2 | 1 | 2 |
| Summer oxidant episodes | 2 | 2 | 3 | 1 | 2 |
| Total | 14 | 23 | 30 | 12 | 22 |

Central to the concept of a learning organization is the view that the future, though it cannot be predicted, can be shaped. For almost two decades, the Shell Group planners in London have experimented with various methods, including scenario planning, to foster strategic thinking. Over time, they have learned that what is important is not whether any scenarios come to pass (although the scenarios presented are plausible futures), but how the exercises help managers improve their mental models of reality. Managers are thus better prepared to cope with unexpected change, since the exercise of visiting plausible futures promotes flexible organizational responses.

In *Changing Course*, S. Schmidheiny (ed.) (1992) MIT Press, Cambridge, MA, p. 93.

summer oxidant episodes. If present socioeconomic trends were to continue or strengthen, nearly all of the issues listed in the Table would become important by the year 2030. If Europe and the rest of the world were to adopt environmentally pathways, these problems could be avoided. But if only Europe were to take this route, it is not likely to escape sea level rise, shortages of freshwater and timber, and other globally generated unpleasant problems.

In the policy-exercise approach, the objective is not to predict the future, but to seek policies that will be robust, no matter what the future may bring – including the possibility of some rather unpleasant surprises. Clearly, a policy exercise is only a single step in the continuing dialogue that ought to take place with respect to long-term policy formulation between scientists and policy people.

## Selected References

Benedick, R. E. (1993) Creating a spirit of partnership: policy dialogues and environmental diplomacy, Capacity Building, International Academy of the Environment, Geneva, Switzerland, pp. 4–6.

EC (1996) *Proceedings, Workshop on Integrated Environmental Assessment*, EC Directorate XII, Brussels, Belgium (In press).

Munn, R. E. (ed.) (1994) *Looking Ahead: The Inclusion of Long-Term 'Global' Futures in Cumulative Environmental Assessments*. CEA 9129, G31. Canadian Electrical Association, Montreal, Canada, 286 pp.

Munn, R. E. (ed.) (1995) *Atmospheric Change in Canada: Assessing the Whole as well as the Parts*. Report of a Workshop, Institute for Environmental Studies, University of Toronto, Toronto M5S 1A4, 36 pp.

Ontario Forest Policy Panel (1993) *Diversity: Forests, People, Communities*, Queen's Printer for Ontario, Toronto, Canada, 147 pp.

SCOPE 5 (1979) *Environmental Impact Assessment*, 2nd edn (R. E. Munn, ed.), John Wiley, Chichester, 190 pp.

Stigliani, W. M., Brouwer, F. M., Munn, R. E., Shaw, R. W. and Antonofsky, M. (1989) Future environments for Europe: some implications of alternative development paths, *Sci. Total Env.*, **80**, 1–102.

# Appendix 5. Inventory of Adaptation Strategies Relating to Climate Change: A Canadian Example[1]

This Appendix contains the results of a literature search for specific adaptations to climate change and variability that could be made in several sectors and geographic settings. The search was widespread in order to discern as many adaptive actions as possible in each sector. However, even this inventory is not necessarily exhaustive. Rather, it demonstrates the huge variety of forms of adaptation which have already been noted in published material. Most of these are primarily concerned with long term climatic change, and they tend to reflect conventional approaches to adaptation. Yet many also have direct relevance to current variability in climate, regardless of climate change.

The actions listed here show that 'adaptation' ranges from basic research to making markets more competitive. Many actions could be characterized simply as making more efficient use of natural resources. The list has not been evaluated in any way; it is simply an inventory of possible adaptations from published sources.

## Agriculture

*Adaptive responses in the first five categories facilitate production under changing climatic conditions. These actions would be undertaken by individual farmers, although government agencies and others might provide information and, where necessary, incentives. These actions are long-term and strategic.*

### *Change Land Topography*

Reduce runoff and improve water uptake:
- subdivide large fields to reduce water avalanching
- grass waterways

[1]From *Adaptation to Climatic Variability and Change*, 1993 (B. Smit, ed.) Occasional Paper 19, Department of Geography, University of Guelph, Guelph, Canada
Prepared by Deborah Herbert, Environmental Adaptation Research Group, Atmospheric Environment Services, Downsview, Canada.

*R. E. Munn, J. W. M. la Rivière and N. van Lookeren Campagne (eds), Policy Making in an Era of Global Environmental Change, 197–218.* 

- land levelling (in order to spread water over fields and increase time of infiltration, reduce runoff, and allow drainage at non-erosive velocities)
- waterway-levelled pans: combination of land levelling, detention dykes, and spillways constructed in shallow, meandering natural water-ways to intercept, store, and use the runoff that normally flows through them
- bench terracing
- tied ridges
- lagoon enlargement (level out existing shallow lagoons and use for cropping: semi-controlled natural irrigation)
- deep ploughing: one-shot treatment to dilute and break up thin layers of impervious clay and hardpan near the soil surface (shallow in the soil profile) to improve water uptake and reduce evaporative losses

Reduce wind erosion:
- produce or bring to soil surface, aggregates or hard clods large enough (1–7 cm in diameter) to resist wind force
- roughen land surface to reduce windspeed impact by farm implement furrows
- spray flat soil surfaces or harden soil clods to reduce evaporation

*Introduce Artificial Systems onto Farm to Improve Water Use and Availability and to Protect against Loss of Soil through Erosion*

Introduce irrigation or change to more efficient irrigation method
- improve efficiency of existing irrigation system:
- dormant season irrigation: use 5–10 tonnes/ha crop residue, irrigate to fill root zone during cool non-crop dormant season (ground water irrigation only)
- line canals or install pipes to reduce conveyance losses
- use brackish water to irrigate where possible
- apply water only when it is used most effectively by plants
- concentrate irrigation water during peak growth period for best crop response:
- corn: water should be concentrated during tasselling time and again during early grain filling
- grain sorghum: before and shortly after head emergence from the boot
- winter wheat: boot stage to soft dough
- millet: only 5 cm supplemental water during head emergence dramatically boosts yields
- on level fields, recycle tailwater, irrigate alternate furrows to improve water-use efficiency of gravity irrigation systems
- reduce amount of water per irrigation
- switch to crops that use less water
- install a more efficient irrigation system
- use sprinkler irrigation systems:

- centre-pivot systems are good for undulating terrain, but seldom have water-use efficiencies above 80%
- drip irrigation systems virtually eliminate losses to evaporation, but are expensive and not suitable for field crops.

If precipitation decreases in a farm area, import water for irrigation from areas where water is more plentiful.

Use diversions.

Use microslopes between wide rows and microwater runoff shields of metal, cement, butyl rubber, or plywood.

Use vegetative barriers and/or snow fences to catch snow to increase soil moisture.

Use windbreaks to protect soil from wind erosion.

Water harvesting: covering a site with impervious membranes or treating water chemically or mechanically to induce runoff.

Applying black polyethylene films between plant rows eliminates evaporation and warms up surface soils, which speeds up plant growth and generates more available soil nitrogen by accelerating organic matter decomposition.

Use drainage systems if precipitation becomes excessive.

Install water points for livestock if area becomes drier.

For livestock:

- improve rations using supplementary protein, vitamins, and minerals
- reduce stocking rates
- wider use of feed conservation techniques and fodder banks.

*Change Farming Practices*

*Actions in this category include various types of cropping practices and other strategies to conserve soil moisture and nutrients, reduce runoff, and control soil erosion.*

*Cropping Alternatives*

- conventional bare fallow
- stubble mulching
- use straw mulches to soften raindrop impact and speed water intake
- minimum tillage
- no-till farming
- plant furrows on the contour to reduce runoff
- reduce field width by strip cropping to control wind erosion
- use legume and sod-based rotations to conserve soil and water
- use wider row spacing to prolong root extension into untapped stored soil water: delays drought stress until rains arrive

- contour cropping to slope: drill furrows can act as little dams for water or snow catchment
- deep furrow drilling for small grains provides a better microclimate, reducing winter kill and protecting plant crowns from wind and soil blast
- establish and maintain vegetative cover to protect the soil against wind and water erosion
- avoid monocropping
- practise crop rotation
- until new varieties can be bred and tested, alternative cropping patterns should be explored; measures of this type can be investigated with minimal expenditures and without major new genetic research

*Other Strategies*

- chisel up soil clods to act as miniature dams for water
- reduce crop populations to match water supply, thus assuring some grain production
- use lower planting densities
- applying modest amounts of fertilizer is the cheapest and most profitable way of increasing crop water-use efficiency
- vary fertilizer application
- use more efficient fertilizer and pest management to preserve water and soil
- use pre-emergence herbicides for summer crops like millet and wide-row, low populations of corn and sorghum; each kilogram of weed tissue reduces yield of dryland crops about 2 kg whether the weed grows during the dormant season or during the crop season
- cease to cultivate land that becomes too arid due to mid-continental drying or in steep, highly erodible areas
- increase production on good land to remove pressure on marginal lands

*Change Timing of Farm Operations*

Change dates of seeding to better fit new climate regimen:

- earlier seedings of new varieties of millet, corn, and sorghum
- autumn seedings of winter wheat in late September instead of late August may be desirable to delay soil water exhaustion until spring rains arrive

More tillage and planting operations in the drier autumn season could offset potential difficulty in getting onto the fields with equipment during the wetter spring.

Plant earlier to offset moisture stress during a warmer and possibly drier summer.

*Use Different Crops or Varieties*

Match crops with water supply and temperature:

- sugar beets and alfalfa:
  - long growing season, requiring high seasonal water demand
  - poorly suited for dryland
  - alfalfa: max. water demand 60 cm
- fall-planted small grains (winter rye, winter wheat, winter barley)
  - have peak demand from mid-May to mid-June
  - can take advantage of stored soil water accumulated in fallow to carry them over a low demand winter succeeded by peak spring rainfall to match peak growth demands
- winter wheat: max. water demand 40–45 cm
- corn and sorghum
  - have peak demands in August
  - very little dryland corn or grain sorghum is grown when annual precipitation is less than 48 cm year except on sandy soil with high water intake capacity
  - corn and grain sorghum: max. water demand is 50–55 cm
- proso millet
  - short season crop
  - max. water demand 30–35 cm
  - in many of the cooler, drier semiarid regions of the world, proso millet would be a logical substitute for barley or wheat in upgrading food production per unit of water available.

Switch from spring wheat to winter wheat in southern Prairies to avoid summer stress.
Adopt new crop varieties and cultural practices to inhibit pests and diseases.

*Develop Governmental and Institutional Policies and Programmes to Reduce Impacts*

*These adaptive actions are undertaken by government and are purposeful and strategic, although they can be of either short or long duration.*

Modify price support and other government programmes to encourage farmers to react quickly to climate change; these programmes could include:
- crop insurance
- commodity co-ops and marketing boards
- stabilization programmes and subsidies
- tariffs and other trade barriers

If climate change seriously undermines the viability of the farm sector:
- support a declining or inefficient agricultural sector only if there is a need to meet other social objectives, such as food security and preservation of the rural community.
- Induce farmers to leave agriculture by:
  - maintaining macroeconomic policies to assure that human and other resources moving out of agriculture can find employment elsewhere in the economy

- develop policies and institutions which facilitate movement between sectors
- create positive incentives for people to leave agriculture:
- stable economic growth
- provide education, specialized skills necessary in modern world.

Nations with declining agricultural production should allow more importation from countries with improved production to avoid higher costs.

Improve storage in areas where yields tend to be unstable.

Take marginal lands out of production and encourage production in most efficient areas:

- discourage temporary colonization of marginal areas
- protect arable land from encroachment by competing uses.

If agriculture shifts in location, new farm-to-market transport links might be needed; rights-of-way might be purchased now before land appreciates

Prepare pest infestation programmes for possible northward shifts in pest locations.

Develop gene, germ plasm reserves.

*Promote Research to Mitigate or Prevent Potential Impacts*

Use genetic engineering to increase plant tolerance to a larger variety of conditions without sacrificing yield.

Develop drought-resistant plants:

- e.g. lower effective wilting point of wheat by developing varieties that can extract more water from the soil
- e.g. reduce leaf size or numbers to lower transpiration (but without also lowering rate of photosynthesis).

Develop heat-resistant plants.

Develop salt-tolerant crops.

Develop food products that can be more easily stored.

Maintain adequate supply of heat and drought-resistant crops in reserve.

Develop new cultivars, cultural practices, and machinery, all specifically adapted to changing climate conditions.

Promote biocontrol and the plant rhizosphere, one goal being to uncover new, beneficial uses for insects, nematodes, and microorganisms.

Improve desalinization technology to reduce its costs.

## The Arctic

*Adaptations to climate change in the Arctic fall into two categories: protection of human beings and their activities from the natural environment, and protection of the natural environment itself. In both cases the actions are strategic and purposeful and would likely be undertaken by governments at the municipal, regional, national, and international levels, depending on the scope of the action.*

*Protect Humans and their Activities from Natural Environment Hazards*

Drainage control and river basin management
Storage of:
- food
- fuel
- public information
- heritage funds.

Iceberg management
- bury installations beneath ocean floor
- construct fixed berms that will stop or deflect all icebergs with draft greater than the top of the berm
- towing, pushing icebergs
- disintegration of icebergs with explosives
- icebergs may be source of freshwater for arid areas or where salinity increases.

Institute buffers against loss:
- drought relief
- international aid
- frost insurance
- welfare.

'Climate-smart' town and transportation planning, incorporating potential for future climate change.
Examine implications of reduced sea ice with respect to:
- icebreaker support
- human settlement development

Create new methods for financing the investment necessary for successful adaptation.

*Protect the Natural Environment*

Stock wildlife reserves and manage forests to achieve optimum natural productivity.
Policies for wildlife protection should focus on human activities and seek to prevent destruction of habitat.
Protection of fertile 'oases', estuaries, and polynyas.

## Coastal Areas

*Protection*

*Measures in this category seek to protect existing structures and activities from sea level rise. These are short or long-term buffering actions; depending on the size and nature of the area to be protected, protection measures can be undertaken by government or by individuals. In the lists of hard and soft protection measures below, notation is used to indicate what each measure is used for, as follows: E = control of erosion, I = protection from inundation, S = protection from storm surges*

*Hard Protection Measures (i.e., those involving physical structures)*
Often have negative effects on adjacent beaches and have high maintenance costs.

- Offshore breakwaters (E)
  - used for beach protection
  - built parallel to beach in fairly shallow water depths
  - designed to intercept most of incident wave energy and decrease transport capacity of waves
  - capital costs can be quite high, especially on open coasts having higher waves
- Perched beach (E, I)
  - continuous, well-submerged offshore structure is built parallel to shore
  - space between shore and structure is filled with sand to create a beach
  - costs between one-quarter and one-half as much as breakwaters per unit length of structure
  - costs one-half to three-quarters as much as breakwaters per unit length of beach
- Groins (E)
  - built perpendicular to beach to trap sand transported by longshore waves or to keep existing sand from being transported away
  - have little effect on sand transport during storms unless waves are extremely oblique
  - cost per unit of structure is usually less than that of breakwaters
- Revetments (E, I, S)
  - structure typically consisting of loose armour material, stones, and concrete blocks laid on a relatively flat slope to protect an embankment from wave attack
  - used where there is little or no beach and a low to moderate wave climate
  - rarely used on open ocean coastlines
- Dykes, levees (E, I, S)
  - earth-filled mound, usually trapezoidal (base is about five times length of top), placed at land-sea boundary to protect lower land area from flooding
  - where there is enough room for it and where fill can be found, a dyke is the best way to control flooding, except where there is exposure to strong wave action

  - dykes can easily be modified to accommodate sea level rise by adding fill to the top and backside and by extending the revetment
  - if the land protected by the dyke becomes lower than the sea, a canal/pump system might be necessary to remove water that seeps into the area, as well as runoff from precipitation
  - construct locks to connect interior navigation channels with the sea
  - may consider turning lands far below sea level into freshwater lakes for fisheries and recreation
  - levees are built for conducting rivers to the sea
- Floodwalls (I, S)
  - used in urban areas where there is not enough land for dykes
  - usually made of concrete
- Seawalls (E, I, S)
  - use in areas of extreme wave action where shore erosion and inundation are to be completely controlled
  - usually made of concrete
  - use less space than dykes or levees
  - cost much more than revetments or bulkheads, so would be economic only where wave action is severe or to protect valuable property
  - developed areas can be protected by seawalls as it becomes necessary
- Bulkheads (E, I, S)
  - vertical wall at land–water boundary to retain fill
  - usually constructed where strong wave and current action are unlikely, eg. marinas, harbours, and along inland waterways
  - good for protecting land in sheltered and semi-sheltered bays and estuaries from sea level rise
- Dams/tidal barriers (I, S)
  - concrete or earth-filled structure placed at the mouth of an estuary or tidal river to prevent storm surges from going upstream
  - have gates which can be opened to allow navigation
  - usually used in conjunction with levees.

*Soft Protection Measures:*

- Artificial beach nourishment (E, I, S)
  - can be used to stabilize some shore protection structures like bulkheads and dykes
  - feasible only where good source of sand is close to area to be nourished
  - offshore deposits
  - deposits at the ebb and flood deltas of a tidal inlet
  - onshore or in nearshore embayments
  - nourishing a beach to maintain a pre-sea level rise location steepens the beach face, making it more prone to erosion and more in need of stabilization by structures

  - many beaches are stabilized by groins or offshore breakwaters to minimize renourishment requirements
- Dune building (E, I, S)
  - a line of continuous dunes just landward of the beach can limit storm inundation (acts like a dyke) and beach erosion (provides a supply of sand)
  - can build dunes by:
  - mechanically placing sand
  - slowly trapping wind-blown sand with fences or vegetation
- Marsh building (E, I, S)
  - salt water marshes can protect coastal areas by absorbing water energy
  - they also trap sediment, and thus the marsh surface slowly grows upward
- Ensure adequate protection of hazardous waste facilities, including
  - landfills
  - surface impoundments
  - land treatment
  - waste piles
- Develop a lead agency for coastal defence (against sea level rise) to:
  - aid information flow
  - provide a base for quickly launching emergency coastal defence efforts

*Allow for Landward Migration of Wetlands*

*Some or all wetlands will be threatened by sea level rise unless they are permitted to move landwards. Because wetlands provide valuable services to humans as well as habitat for many species, it is important to preserve as many wetlands as possible. Although the majority of actions in this category would likely be undertaken by government, individuals should also endeavour where possible to ensure the survival of wetlands. These actions are strategic and long term.*

Prohibit construction of bulkheads.
Restrict coastal development by adopting set-back requirements:
- but if sea level rise is unknown, then so too is the proper set-back requirement
- a less costly alternative is the 'presumed mobility' approach adopted by the state of Maine: property owners assume responsibility for moving structures as sea level rises

Sea level rise must be incorporated into environmental and land ownership policies or else wetlands will be squeezed between sea level and land development.

Government could exercise eminent domain and pay landowners full price to allow wetland migration, but this could be very expensive, unless land is purchased before it is developed (i.e. in anticipation of sea level rise and wetland migration).

Make institutional changes:

- laws that define ownership on the basis of tidal regimens could be amended with the elaboration that if tides change, then so too does land ownership
- include provision in deeds for coastal lands that land would revert to public ownership in so many years (e.g. 100) if sea level rises some specified amount
- allow development of these areas in exchange for agreement to vacate the land if sea level rises.

*Water Supply and Sanitation*

*Sea level rise may threaten coastal freshwater supply as well as coastal drainage systems. Even land safe from inundation will be virtually uninhabitable if there is no longer a source of fresh water or drainage becomes impossible. Actions to protect freshwater supply and ensure proper drainage would likely be undertaken by whoever has responsibility for provision of these services: local government, privately-owned companies, or individuals. These actions are strategic and long term.*

Responses to saltwater intrusion:

- Physical subsurface barriers:
  - driving sheet pile, installing a clay fence, or injecting impermeable materials through wells
  - very expensive because the required depths are substantial
  - backwater effect could cause waterlogging of coastal lowlands
- Extraction barriers:
  - collection and removal of saltwater that moves inland in aquifers
  - the pumping encourages further intrusion and may inadvertently withdraw freshwater, and thus this would not be an appropriate response in areas where freshwater is scarce
  - generally more expensive than injection barriers; would be economical for protection of small areas like hazardous waste sites or coastal aquifers with narrow connections to the sea
- Freshwater injection barriers:
  - inject freshwater from another area into aquifer, raising the freshwater level and reversing the saltwater intrusion
  - this can be accomplished with a line of several wells along the coastline
- Increased recharge of coastal aquifers
  - spread water on upland recharge areas during high precipitation periods
  - problems could be:
    - lack of sufficient recharge water
    - lack of inexpensive land for recharge basins or shallow injection wells
    - costly technical problems of maintaining a proper inflow rate
- Expand water collection and pumping systems.

Coastal aquifers:

- modify pumping patterns
- reduce withdrawals
- pump further inland
- direct surface water delivery to replace groundwater in use

Coastal drainage systems:

- enhance gravity drainage systems:
  - use larger pipes or install supplemental pipe systems
  - widen and deepen existing drainage canals
- reduce hydraulic roughness of canals
  - eg. line canals with asphalt or concrete
- areas that will no longer be above sea level cannot continue to use gravity drainage and will have to switch to a forced drainage system
- the use of locks and gates (which open to permit gravity drainage in low tides and close in high tides) may be a cost-effective interim solution
- enhance forced drainage systems:
  - larger pumps for new systems
  - additional pumps for old systems
- may also be necessary to increase the capacity of the system that delivers the storm water to the pumping station
- use detention measures to reduce peak discharge:
  - basins (not feasible in highly developed areas where land is valuable)
  - rooftop detention
  - infiltration trenches
  - porous pavement
  - storage in low playgrounds and parking lots
  - in-line storage in storm sewage pipes.

*Elevation/Floodproofing*

*Adaptive responses in this category are typically long-term and strategic. They would occur at the regional level and be undertaken by government (with the exception of elevation of privately-owned buildings and infrastructure, although this could be required by law).*

Elevate structures:

- land height can be built up by placing landfill and holding it with a retaining structure if the area is not too big
- raise marina and harbour infrastructure (e.g. fender systems, docks, walkways)
- raise and protect threatened buildings.

Components like tanks, containers, incinerators, and structures used in hazardous waste treatment (thermal, chemical, physical, biological) can be floodproofed:

- grading
  - create 2–5% slope for runoff
- fencing
  - install around facilities to restrict the flotation of containers and reduce structural damage from flood debris
- upgrade the structural integrity of containers
- elevate containers and/or facilities above flood level:
  - earth fill (provided it doesn't erode)
  - column piers or walls (provided they don't impede floodwaters and are protected against scour).

Amend building codes and engineering standards, as well as zoning and building permit processes, to ensure that sea level rise is taken into account in new construction.
Avoid construction of basements.

*Policies to Ensure that Land Development Incorporates Potential Sea Level Rise*

*These policies would ensure that land vulnerable to inundation from sea level rise is not developed. They would be adopted by government with jurisdiction over the area in question (most likely local government) and are strategic and long-term.*

As appropriate, eliminate government subsidies which encourage development of vulnerable areas.

- zone to prevent/restrict development of vulnerable areas.
- site new hazardous waste facilities where flood damage is minimized.
- raise flood insurance premiums to discourage development in vulnerable areas.

## Ecosystems and Land Use

*Actions in the following two categories seek to preserve ecosystems, either through human intervention or through expansion of the ecosystems. These actions are not specific to climate change adaptation, but they do become even more necessary in the face of climate change. These actions would be initiated and/or carried out by government, although it is important for individuals to undertake these actions on any scale possible if they can. These responses are strategic and medium to long term.*

*Protection of Existing Habitats and Species*

Strategies should aim to protect species at the bottom of the food chain.

Seeds and seedlings should be transplanted to appropriate locations: this would require well-developed techniques of plant propagation, and thus would be suitable only for commercially important species

Adopt 'safe minimum standard, the survival of species, habitats, and ecosystems'.

Expand seed banks and gene pool protection in zoos and reserves.

Abandon strategy of crisis management on a species-by-species basis in favour of overall habitat protection.

*Enlarge Protected Wildlife Habitats to Ensure Species Survival*

Expand size of wildlife refuges to reduce their vulnerability to climate change.

Establish reserves:

- place greater emphasis on climate criteria for designating where parks and other reserves might be needed

Establish migration corridors between reserves, e.g. greenways, hedgerows

Purchase land:

- where a reservoir might be needed in future
- where wildlife might migrate.

Restrict land use before land is developed:

- avoids compensation of developers/users and their efforts to block the restriction.

Means of restricting land use:

- zoning
- direct purchase of land
- set up social constraint
  - e.g. 'ecosystem must be allowed to migrate' regulation
  - e.g. 'presumed mobility':
  - allows development of land, provided that development does not block ecosystem or wildlife migration
  - imposes no costs in face of uncertainty
  - may, however, be a lot of pressure to repeal if climate change does force migration and developers/owners must give up the land
- modify conventions of property ownership
  - leases that expire on some specific date or once some condition occurs

## Energy Supply

*Responses to climate change in this category reduce the vulnerability of power plants to decreased water supply and/or increased fluctuation in water supply levels. The actions are long-term and strategic; they involve in most cases considerable changes to the capi-*

*tal stock or operating procedures of power plants. They would be undertaken by the company or agency which runs the plant or utility, although there is a role for government in providing information and incentives where necessary. There is also a role for government in ensuring that water prices accurately reflect the scarcity of water.*

*Thermal Plants (Fossil Fuel or Nuclear)*

Reduce water withdrawal by using a closed (rather than a once-through) cooling system, where the water withdrawn is reused several times

Use alternative cooling systems:

- dry cooling towers (thermal efficiency is lower, though)
- wet-dry cooling systems
- ammonia-based cooling systems

Water savings do not currently justify the additional capital and operating costs of these alternatives in most places; this could change if climate change reduces water availability.

Conserve water used for cooling (consumption = 2% of water withdrawn).

Use saline water (in coastal areas) or municipal wastewater for once-through cooling.

Improve thermal efficiency (but limited possibilities; capital costs begin to rise sharply as efficiency increases).

*Hydroelectric Plants*

Change the operation of hydroelectric plants:

- invest in water storage
- plan for altered potentials and design factors
- hold excess generating capacity in all but 'critical period' water conditions.

Augment hydroelectric production with energy from other sources:

- operate both hydro and thermal plants
- use higher-cost reserve plants
- import power from other utilities; develop regional power market; long distance power exchange – integrate utilities on national power grid

Change the way hydroelectricity is sold in response to reduced or more variable production:

- market electricity as 'firm' (high price) and 'interruptible' (low price)
- load shedding (generally the interruptible contracts)
- peak-load shifting
- drought surcharges

- voluntary conservation programmes.

## Fisheries

*Measures are strategic and long-term and most are undertaken by government at the regional, national, and perhaps international levels, although some can be undertaken by individuals or the private sector (e.g. aquaculture).*

### *Make Fisheries Sustainable in the Long Run and Under a Changing Climate*

Modify fishery operation and strengthen fish population monitoring to prevent overfishing and ensure sustainable fisheries.
Consider fishery and wildlife habitat needs in planning coastal defence measures and in pollution control programs.
Fish breeding, supported by measures to preserve genetic diversity of fish populations.
Restock with ecologically sound species.
Experimentally stress marine systems to discover the boundaries of stability for different states.
Encourage novelty in fishery operations

### *Develop Alternatives to Traditional Uses of Fishery Resources*

Develop aquaculture.
Create employment outside the fishing sector.
Explore use of aquatic and marine species as indicators of change.

## Forestry

### *Protect Existing Forests*

*Responses in this category are strategic and long-term and would most likely be undertaken by government, although universities and the private sector would also be useful participants in projects.*

Use integrated biosphere observatories/reserves as holistic monitoring and early warning areas for impacts.

Enhanced protection of forests, especially in terms of prevention:
- education programmes aimed at fire prevention
- enhanced fire and pest monitoring and fire fighting efforts
- take stock of actual and potential biological invasions
- create training centres for state-of-the-art technologies for suppression of the impacts of fire, insects, diseases, frost, drought, and so on

Consider the impacts on non-commercial forest values of changes in forest management practices

*Because the majority of actions in the following two categories would provide benefits to the private sector, it may undertake these actions on its own. However, these actions are mostly long-term so where planning horizons are short, the benefits may not be realized or recognized by forest companies, implying that national or regional governments would have to provide incentives. There would also be a role for government in disseminating information to foresters.*

### *Introduce New Species Where Appropriate*

Introduce new species

Potential species for boreal forest region:
- native species;
- trembling aspen
- lodgepole pine and black pine
- black spruce and white spruce
- tamarack

Exotic species:
- red pine
- green ash
- oaks and pines

Breed traits such as disease and drought resistance in trees

Conserve gene pools:
- seed banks

### *Change or Improve Use of Forests*

Partition forests to more clearly set apart managed areas.

Use shorter rotation times for commercial forests:
- then replace with adaptable species where possible
- but species that will do well under climate change conditions may be difficult to establish in today's climate.

Harvesting currently occurs during winter when frozen ground allows easy access; this may change due to climate change: implications should be explored and appropriate changes made.
Harvesting should leave a diversity of species to enhance regrowth.
Review nursery operations in light of potential impacts of climate change.
Explore use of plantation forests as a source of fuel and to extend natural range of boreal forest.
Develop alternative products.
Concentrate management efforts on sites not likely to be negatively affected by climate change:

- moist areas in southern region of forest zone
- central-northern region of boreal forest.

Move forest industries closer to new sources of supply if forest boundaries shift.

## Urban Infrastructure

*The actions listed below ensure that urban infrastructure is designed to accommodate potential climate changes. They are strategic, mostly long-term actions to reduce the vulnerability of cities to a changing climate. Most actions would be undertaken by municipal governments with the cooperation of trades associations and standard-setting agencies.*

Revise standards and safety factors for the following as they may be affected by climate change:

- ventilation
- drainage
- flood protection
- facility siting
- expansion capability
- resistance to corrosion

Adjust geographically based standards to reflect new conditions, e.g.

- roadbed depth
- home insulation.

Redesign water cooling towers to reduce evaporation losses.

## Water Resources

*The actions listed below are strategic and long-term. They would be carried out by water supply agencies, but with the cooperation of and, where necessary, encouragement by government (where the supply agency is not a government department), at primarily a municipal or regional level, although some regulations could be instituted at the national or international level.*

*Alter Use of Water to Better Reflect Changing Water Flows under a Changing Climate*

*Structural Measures*

- consider climate change in planning and design of new projects
- determine how water flows (effluent discharge permits may need to be changed)
- projects should be phased to maximize flexibility
- increase water storage capacity
  - however, in many cases, it would be more cost-effective to protect a vegetated watershed than to build more reservoir storage
- reduce losses to evaporation and seepage
- increase size of canals, pipelines, and pumping plants to accommodate increasing variability in runoff that may accompany climate change
- relocate intake structures to accommodate river channel changes
- increase control of flow
- develop and commercialize technologies that reduce the cost of desalinization and water recycling

*Operational Plan Changes*

- use sensitivity analysis and risk-cost analysis to incorporate uncertainty of climate change parameter predictions into project and operations planning
- directly tie system operating criteria (eg. rule curves) to climatic indices to ensure automatic adjustment to climate change (the rule curve itself would have to be replaced periodically to reflect changing relationships, etc.)
- perhaps, give recent events/trends more weight than earlier ones in statistical analysis
- make changes in policy, before changes occur, to minimize opposition from potential losers
- conjunctive use of surface and groundwaters:
  - surplus winter runoff that cannot be stored in already full reservoir is artificially recharged into groundwater basin for later use during drought or emergencies

*Water Conservation and Demand Management*

Conserving water is a good idea even in the absence of climate change as it reduces need for expensive new water projects.

Demand management by allocation plans.

Educate the public on need to reduce water; education programmes have longer-lasting effect than programmes which appeal to consumers' goodwill.

Rewards might also be used to encourage conservation.

Use treated municipal wastewater for lower-quality uses, such as irrigating parks and golf courses.

Trade municipal wastewater with agriculture for higher quality water.

Directly regulating water use can be a useful adjunct of water pricing policies e.g.

- alternate-day lawn watering
- banning use of ornamental fountains.

Demand management could be achieved by:

- education and voluntary compliance
- price policies
- legal restrictions
- rationing
- imposing water conservation standards on things like appliances and showerheads

*Improve the Efficiency of Water Use Through Management Plans*

Through drought management plans:

  - to be in place before the changes occur (i.e. before droughts become more frequent or more severe)
  - set priorities for users ahead of time to minimize conflict
  - water pricing plans:
  - put in place now; to come into effect when some threshold level is reached.

Through the reallocation of water rights

- allow sale of water rights
  - but it may be necessary to let the government oversee the transfer of water rights to protect third party and public interests
- some alternatives to the sale of water rights:
  - leasing and conditional (i.e. upon the occurrence of some event) leasing
  - water banking
  - user with excess water can voluntarily 'deposit' that water in a water bank in exchange for some market-determined price
  - deposits can be temporary, annual, or long-term; they can be withdrawn by other users for a fee
  - limited by scope of water delivery system
  - requires extensive data and bookkeeping
- give irrigators property rights for conserved water and salvaged runoff to give them a financial incentive to conserve water

- change the basis for a water right from withdrawal quantity for a specified use, to consumptive quantity for that use
- requiring that water entailed in aboriginal water rights must be used on the reservation limits the value of that right and reduces the amount of water available for other uses
- remove riparian water rights with prior appropriation (i.e. first in time, first in right) to create definable property rights
- remove restrictions to certain uses (beneficial uses) or locations.

Through satisfaction of economic, rather than social, goals in water supply
- remove subsidies
- price water at its replacement cost and allow markets to allocate water to its most effective uses
- where average cost pricing is used, charge more for successive blocks of water use
- price water at or close to its marginal cost
- if marginal cost pricing is used, water agencies can avoid hurting the poor, and can earn profits in various ways:
  - use a rebate scheme, where profits are redistributed among those who experienced hardship due to the higher water prices
  - use a 'lifeline' scheme, where some initial necessary quantity of water is priced below cost and additional quantities are priced above cost (this option is probably preferable to the rebate scheme because of the red tape and expense of administering rebates.)
  - reduce prices:
    - first, for customer charges unrelated to consumption
    - then, for early blocks of winter usage
    - then, for early blocks of summer usage
    - varying prices among seasons and locations can be useful way to introduce marginal cost pricing
    - divide area serviced into elevation and distance zones and charge customers in each zone a price equal to the cost of supplying water to that zone
    - also permits more effective planning of expansion and inhibits urban sprawl

*Protect Instream Flows and Water Quality*

*Where water is unappropriated, it might be best for government to claim the water right and use it to protect instream flow; the right could later be sold if others need the water*

Instruments for government to protect instream flows:
- appropriate the relevant water rights
- set minimum streamflows beyond which water cannot be removed for offstream uses

- define instream flow as a use open to individual or group appropriation
- legislate protection directly.

Ecosystem needs should be integrated into water resource planning

*Interbasin Transfers*

Increase capacity for interbasin transfers; this may be prudent even in the absence of climate change.

As a backup, managers of adjacent water systems could share water resources as needed during dry periods.

Interbasin transfers have certain costs/constraints:

- environmental costs:
  - lost recreation sites and wildlife habitat in area transferring the water
  - potential for ecological disruption, especially where foreign species are introduced into the area receiving the water
- economic costs are huge, often prohibitive
- people who are transferring water will want compensation, even if they have excess water
- may also cause political conflict
- climate change will complicate planning:
  - long-term droughts tend to be localized but climate changes will be persistent and widespread
  - uncertainties associated with predictions of regional impacts may take decades to resolve.

Treaties for sharing of international water resources (i.e. rivers) should specify:

- quality of water to be delivered
- clear definitions of 'climate change' and 'drought', as well as how to identify the onset of such events
- unambiguous allocations of shortages.

# Index




*R. E. Munn, J. W. M. la Rivière and N. van Lookeren Campagne (eds), Policy Making in an Era of Global Environmental Change, 219–225.*